Authored by Hemen Dutta, Pinnangudi N. Natarajan, and Yeol Je Cho

Concise Introduction to Basic Real Analysis

Authored by Hemen Dutta, Pinnangudi N. Natarajan, and Yeol Je Cho

Concise Introduction to Basic Real Analysis

CRC Press
Taylor & Francis Group
Boca Raton London New York

CRC Press is an imprint of the
Taylor & Francis Group, an **informa** business

CRC Press
Taylor & Francis Group
6000 Broken Sound Parkway NW, Suite 300
Boca Raton, FL 33487-2742

First issued in paperback 2020

ISBN-13: 978-1-138-61246-4 (hbk)
ISBN-13: 978-0-367-77928-3 (pbk)

Library of Congress Cataloging-in-Publication Data

Names: Dutta, Hemen, 1981- author. | Natarajan, Pinnangudi N., author. | Cho, Yeol Je, author
Title: Concise introduction to basic real analysis / by H. Dutta, P.N. Natarajan, and Y.J. Cho.
Description: Boca Raton : CRC Press, Taylor & Francis Group, 2019.
Identifiers: LCCN 2019007146 | ISBN 9781138612464 (hardback : alk. paper) | ISBN 9780429464676 (e-book)
Subjects: LCSH: Functions of real variables–Textbooks. | Mathematical analysis–Textbooks. | Numbers, Real–Textbooks.
Classification: LCC QA331.5 .D88 2019 | DDC 515/.88–dc23
LC record available at https://lccn.loc.gov/2019007146

Visit the Taylor & Francis Web site at
http://www.taylorandfrancis.com

and the CRC Press Web site at
http://www.crcpress.com

Contents

Preface

The contents and approach to the subject Real Analysis adopted in this book is expected to be useful for undergraduate and graduate students in understanding the basic concepts. The book attempts to provide a reasonable introduction to basic topics in real analysis and to make the subject digestible to learners. Readers will find solved examples and chapter-end exercise in each chapter including hints for solution. The book contains nine chapters and is organized as follows.

In the chapter "**Review of Set Theory**", we introduce basic concepts of set theory. We define Cartesian product, relations and functions, countable and uncountable sets. We prove that the set of all real numbers is uncountable and the set of all rational numbers is countable. We then introduce the set algebra, thereby defining operations on sets such as union, intersection, and complement, and prove their properties.

In the chapter "**The Real Number System**", we discuss the real number system. We take real numbers for granted satisfying certain axioms from which further properties are derived. In this direction, we introduce the field axioms, the order axioms, and the completeness axiom. We then proceed to derive several useful properties of real numbers.

The chapter "**Sequences and Series of Real Numbers**" is devoted to a systematic study of sequences and infinite series of real numbers. In the context of real numbers, we introduce the concepts of Cauchy sequences, limit superior and limit inferior, convergent and divergent sequences, convergent and divergent series, absolutely convergent and conditionally convergent series and study their properties. We then prove some convergence tests for infinite series. We further discuss rearrangement of infinite series and prove Riemann's theorem on conditionally convergent series. Finally, we discuss the Cauchy multiplication of infinite series.

In the chapter "**Metric Spaces – Basic Concepts, Complete Metric Spaces**", we introduce metric spaces and related basic concepts. We then discuss complete metric spaces. We prove the Bolzano–Weierstrass theorem and the Cantor intersection theorem. We then introduce Cauchy sequences, convergent and divergent sequences in a metric space and discuss their properties in detail.

In the chapter "**Limits and Continuities**", we first discuss limit of a function, right-hand limit, left-hand limit, infinite limits, limits at infinity, sequential definition of limit, Cauchy's criterion for finite limits and present several results covering various properties of limits with examples. Next we discuss monotonic functions, continuous and discontinuous functions by giving several related results with examples. Then, the notion of uniform continuity is discussed with results and examples. Finally, we give a brief discussion on continuity and uniform continuity in metric spaces.

The chapter "**Connectedness and Compactness**" deals with the notion of connectedness in metric space, Intermediate Value Theorem, components, compactness in metric space, Finite Intersection Property, totally bounded sets, the Bolzano–Weierstrass Theorem, sequentially compact spaces, and the Heine–Borel Theorem.

In the chapter "**Differentiation**", we deal with the definition of the derivative, the differentiable functions, and their properties; the chain rule; derivative of inverse functions; Rolle's theorem; Lagrange's mean value theorem; Intermediate Value Theorem for derivative; Cauchy's mean value theorem; indeterminate forms; L'Hospital rule; Taylor's theorem; Taylor series; Maclaurin series; and the local minimum and maximum.

The chapter "**Integration**" discusses the Riemann integral, examples and properties of Riemann integral, Fundamental Theorems of Calculus, Mean Value Theorems for integrals, Substitution Theorem for integrals, Theorem on Integration by Parts, improper Riemann integrals, the Riemann–Stieltjes integral with examples and properties, and functions of bounded variation.

In the chapter "**Sequences and Series of Functions**", we first introduce pointwise convergence of sequences of functions, uniform convergence and study their properties. We study uniform convergence and continuity, uniform convergence and integration, and uniform convergence and differentiation. Equicontinuous family of functions is then introduced, and we prove the Arzela–Ascoli theorem. Dirichlet's test for uniform convergence and the Weierstrass theorem are also established.

The authors are grateful to several friends and colleagues for their encouragement and mental support while the manuscript was being prepared. The authors must thank several authors and contributors in the field of real analysis for their valuable contribution which directly or indirectly influenced the preparation of this book. Constructive suggestion and criticism shall be taken with full sincerity for possible implementation in future editions.

H. Dutta,
Guwahati, India
Pinnangudi N. Natarajan,
Chennai, India
Yeol Je Cho,
Jinju, Korea

Authors

Hemen Dutta is a faculty member at the Department of Mathematics, Gauhati University, India. He did his Master of Science in Mathematics, Post Graduate Diploma in Computer Application, and Ph.D. in Mathematics from Gauhati University, India. He received his M.Phil. in Mathematics from Madurai Kamaraj University, India. His primary research interest includes areas of mathematical analysis. He has to his credit several research papers, some book chapters, and few books. He has delivered talks at different institutions and organized a number of academic events. He is a member of several mathematical societies. He has also been a resource person in training programs for school students and teachers and delivered popular talks. He has authored several newspaper and popular articles in science education, research, etc. He has served the Assam Academy of Mathematics as joint-secretary (honorary) for two years.

Pinnangudi N. Natarajan is Dr. Radhakrishnan Awardee for the Best Teacher in Mathematics for the year 1990–1991 by the Govt. of Tamil Nadu, India. He has been working as an independent researcher in Mathematics since his retirement in 2004, as professor and head, Dept. of Mathematics, Ramakrishna Mission Vivekananda College, Chennai, India. He received his Ph.D. in Analysis from the University of Madras in 1980. He has to his credit over 100 research papers published in several reputed national and international journals. He has authored three books (one of them in two editions) and contributed a chapter to each of the two edited volumes, published by Springer, Taylor & Francis Group and Wiley. Dr. Natarajan's research interest includes Summability Theory and Functional Analysis, both Classical and Ultrametric. Besides visiting institutes of repute in Canada, France, Holland, and Greece on invitation, he has participated in several international conferences and has chaired sessions.

Yeol Je Cho is an Emeritus Professor at Department of Mathematics Education, Gyeongsang National University, Jinju, Korea, and a Distinguished Professor at School of Mathematical Sciences, University of Electronic Science and Technology of China, Chengdu, Sichuan, China. Also, he is a fellow of the Korean Academy of Science and Technology, Seoul, Korea, since 2006, and a member of several mathematical Societies. He received his Bachelor's Degree, Master's Degree, Ph.D. in Mathematics from Pusan National University, Pusan, Korea, and, in 1988, he was a Post-Doctor fellow in the

Department of Mathematics, Saint Louis University, Saint Louis, USA. He is an organizer of the International Conference on Nonlinear Functional Analysis and Applications, some workshops and symposiums on Nonlinear Analysis and Applications, and member of the editorial board of 10 more international journals of mathematics. He has published 350 more papers and some book chapters, 20 more monographs, and ten more books with CRC, Taylor & Francis Group, Springer, and Nova Science Publishers, New York, USA. His research areas are Nonlinear Analysis and Applications, especially, fixed point theory and applications, some kinds of nonlinear problems, that is, equilibrium problems, variational inequality problems, saddle point problems, optimization problems, inequality theory and applications, and stability of functional equations and applications. He delivered many invited talks at many international conferences on Nonlinear Analysis and Applications, which have been opened in many countries.

1

Review of Set Theory

For the study of any branch of Mathematics, set theory is always very useful. Set theory was developed by Book and Cantor in the latter part of the 19th century. Development of Mathematics in the 20th century was very greatly influenced by set theory. In the present chapter, we introduce some basic concepts of set theory which are very useful in the sequel.

1.1 Introduction and Notations

We do not attempt to define a "set". All of us agree to consider a "set" as undefined. A set can be "described" as a collection of objects considered as a single entity. The objects of this collection are called elements or members of the set. These objects are said to belong to the set or are contained in the set. The set is said to contain its objects or is composed of its objects.

Sets are usually denoted by capital letters $A, B, C, \ldots$ or $X, Y, Z, \ldots$, and elements of the set are denoted by lower-case letters $a, b, c, \ldots$ or $x, y, z, \ldots$. If x is an element of a set X, we write $x \in X$. In such a case, we also say that "x belongs to X". If x does not belong to X, we write $x \notin X$. Often, we denote sets by displaying their elements within braces. For example, the set of all positive integers less than 12 is denoted by $\{1, 2, 3, 4, 5, 6, 7, 8, 9, 10, 11\}$ or by $\{x : x$ is an integer, $0 < x < 12\}$. More generally, if X is the set of all elements x satisfying a property P, we denoted it by

$$X = \{x : x \text{ satisfies } P\}.$$

Given a set, we can form new sets as follows. For instance, let Y be the set of all positive integers less than 12 which are divisible by 3; then, $Y = \{3, 6, 9\}$. Note that Y contains some elements of the set $X = \{x : x$ is an integer, $0 < x < 12\}$. Y is called a subset of X. More generally, given the sets A, B, we say that A is a *subset* of B, written as $A \subseteq B$ (or $B \supseteq A$) if every element of A is also an element of B, i.e., whenever $x \in A$, then $x \in B$ too. If $A \subseteq B$ and $B \subseteq A$, we then say that A and B are *equal*, written as $A = B$. Note that $A = B$ if and only if A and B have the same elements. If A and B are *not equal*, we write $A \neq B$. If A is a subset of B and $A \neq B$, we write $A \subsetneq B$. In such a case, we say that A is a *proper subset* of B.

A set which contains no elements is called an *empty set* or a *null set*, denoted by $\emptyset$. We agree to call $\emptyset$ a subset of every set.

An example of a null set is the set of all real numbers, whose square is negative. The reader should give more examples of null sets. Note that all null sets are equal in the sense described earlier, and so we talk of the unique null set.

1.2 Ordered Pairs and Cartesian Product

We note that $\{a, b\} = \{b, a\}$ as sets. Here the question of the order in which the elements a, b occur does not arise. However, there are situations where the order in which the elements occur is of paramount importance. For example, in analytic geometry of the plane, the point $(1, 2)$ is different from the point $(2, 1)$. Again, if we consider the complex number $a + ib$ as the ordered pair (a, b), then the pair $(3, 4)$ is different from the pair $(4, 3)$. So, if we wish to consider a set of two elements a, b as being ordered, a occurring first from the left and then b, we shall denote it by the symbol (a, b).

Definition 1.2.1. Given two ordered pair of elements, (a, b) and (c, d), we define $(a, b) = (c, d)$ if and only $a = c$ and $b = d$.

Definition 1.2.2. Given two sets A, B, we define their *Cartesian product*, denoted by $A \times B$, by

$$A \times B = \{(a, b) : a \in A, \ b \in B\}.$$

If $\mathbb{R}$ denotes the set of all real numbers and $\mathbb{C}$ denotes the set of all complex numbers, then

$$\mathbb{C} = \mathbb{R} \times \mathbb{R}.$$

1.3 Relations and Functions

Definition 1.3.1. Any set of ordered pairs is called a *relation*. If R is a relation, the set $\{x : (x, y) \in R\}$ is called the *domain* of R, denoted by $d(R)$ and the set $\{y \in (x, y) \in R\}$ is called the *range* of R, denoted by $r(R)$.

Definition 1.3.2. Let R be a relation with domain $d(R)$. The relation R is said to be:

(a) *reflexive* if $(a, a) \in R$ for every $a \in d(R)$;

(b) *symmetric* if $(a, b) \in R$ implies that $(b, a) \in R$;

(c) *transitive* if $(a, b) \in R$ and $(b, c) \in R$ imply that $(a, c) \in R$.

A relation R which is reflexive, symmetric, and transitive is called an *equivalence relation*.

A special kind of relation, known as a function, is defined as follows:

Definition 1.3.3. A *function* f is a set of ordered pairs no two of which have the same first member, i.e., if $(x, y), (x, z) \in f$, then $y = z$.

In other words, f is a function if, for every $x \in d(f)$, there exists a unique y such that $(x, y) \in f$. We usually call y the *value* of f at x and write $y = f(x)$ instead of writing $(x, y) \in f$. y is called the *image* of x under f.

In this context, we note that the functions f, g are *equal* if and only if f, g have the same domain and $f(x) = g(x)$ for every $x \in d(f) = d(g)$.

Given sets X, Y, by a *function* or *mapping* f of X into Y, we mean a definite rule which associates with every element of X a unique element of Y, written as $f : X \to Y$. Note that $d(f) = X$ and $r(f) \subseteq Y$. The set $f(X)$, defined by,

$$f(X) = \{f(x) : x \in X\}$$

is called the *image* of X under f. Note that $f(X) = r(f)$.

Definition 1.3.4. If $f : X \to Y$ and $f(X) = Y$, the mapping f is said to be *onto* or *surjective*. We note that the mapping f is onto if and only if, for any $y \in Y$, there exists $x \in X$ such that $f(x) = y$. A mapping of a set X into itself is sometimes called a *transformation*.

Definition 1.3.5. If $f : X \to Y$, then f is said to be *one-to-one* or *injective* on X if $f(x_1) = f(x_2)$ implies $x_1 = x_2$ for every $x_1, x_2 \in X$.

It now follows that f is one-to-one if and only if distinct elements of X have distinct images in Y.

Definition 1.3.6. Given a relation R, the new relation $\tilde{R}$, defined by

$$\tilde{R} = \{(a, b) : (b, a) \in R\},$$

is called the *converse* or the *reverse* of R.

Definition 1.3.7. Suppose f is a function. Consider $\tilde{f}$. $\tilde{f}$ may or may not be a function. If $\tilde{f}$ is also a function, then $\tilde{f}$ is called the *inverse* of f, denoted by f^{-1}.

Theorem 1.3.1. *If the function f is one-to-one on its domain, then $\tilde{f}$ is also a function.*

Proof. We have to prove that, if $(x, y), (x, z) \in \tilde{f}$, then $y = z$. $(x, y) \in \tilde{f}$ implies that $(y, x) \in f$, i.e., $x = f(y)$. Similarly, $x = f(z)$. So, $f(y) = f(z)$. Since f is one-to-one, $y = z$. Thus, $\tilde{f}$ is a function. This completes the proof.

Given the functions f, g, we can define a new function as follows:

Definition 1.3.8. Let f, g be functions such that $r(f) \subseteq d(g)$. Then, the *composite* or *composition* of g and f, denoted by $g \circ f$, is defined by

$$(g \circ f)(z) = g[f(x)] \text{ for every } x \in d(f).$$

Note that $f(x) \in d(g)$, and so $g[f(x)]$ is defined. We also note that $g \circ f \neq f \circ g$, even when $f \circ g$ is defined. However, the associative law holds, i.e.,

$$h \circ (g \circ f) = (h \circ g) \circ f,$$

whenever both the sides are defined.

Definition 1.3.9. By a *finite sequence* of n terms, we mean a function f whose domain is the set $\{1, 2, \ldots, n\}$.

Note that

$$r(f) = \{f(1), f(2), \ldots, f(n)\}.$$

We usually write $f(k) = f_k$, $k = 1, 2, \ldots, n$, so that

$$r(f) = \{f_1, f_2, \ldots, f_n\}.$$

Elements of $r(f)$ are called the *terms* of the sequence.

Definition 1.3.10. By an *infinite sequence*, we mean a function f whose domain is the set $\{1, 2, 3, \ldots\}$ of all possible integers.

The range of f, i.e., $\{f(1), f(2), f(3), \ldots\}$ is usually written as $\{f_1, f_2, f_3, \ldots\}$. f_n is called the *nth term* of the sequence. If f_n is the *nth* term of a sequence, then the infinite sequence itself is written as $\{f_n\}$.

Definition 1.3.11. Let $s = \{s_n\}$ be an infinite sequence. Let k be a function, whose domain is the set of all positive integers and whose range is a subset of the set of all positive integers. Let

$$k(m) < k(n) \quad \text{whenever} \quad m < n,$$

i.e., k preserves order. Then, $s \circ k$ is defined for every $n = 1, 2, \ldots$ and

$$(s \circ k)(n) = s_{k(n)} \text{ for every } n = 1, 2, \ldots.$$

The composite function $s \circ k$ is called a *subsequence* of s. The subsequence of $\{s_n\}$, whose *nth* term is $s_{k(n)}$ is denoted by $\{s_{k(n)}\}$ or $\{s_{k_n}\}$.

1.4   Countable and Uncountable Sets

Definition 1.4.1. A set A is said to be *equivalent* to a set B, written as $A \sim B$, if there exists an injective mapping of A onto B.

In this case, we say that there is a *one-to-one correspondence* between the elements of A and B. Note that, if $\sim$ satisfies the following:

(a) $A \sim A$;

(b) if $A \sim B$, then $B \sim A$;

(c) if $A \sim B$ and $B \sim C$, then $A \sim C$,

where $\sim$ is an *equivalence relation*.

Definition 1.4.2. A set A is said to be *finite* with n elements if

$$A \sim \{1, 2, \ldots, n\}.$$

The integer n is called the *cardinal number* of A. The empty set is considered finite with cardinal number 0. Sets, which are not finite, are called the *infinite sets*.

Note that an infinite set is equivalent to a proper subset of itself. For instance, the set $\mathbb{Z}^+$ of all positive integers is equivalent to the proper subset $\{2, 2^2, 2^3, \ldots\}$. The one-to-one function f which makes them equivalent is defined by

$$f(x) = 2^x \quad \text{for every} \quad x \in \mathbb{Z}^+.$$

Definition 1.4.3. A set X is said to be *countably infinite* if it is equivalent to the set of all positive integers, i.e.,

$$X \sim \{1, 2, 3, \ldots\}.$$

That is, there exists a one-to-one correspondence between the sets X and $\mathbb{Z}^+$.

In such a case, we can write

$$X = \{x_1, x_2, x_3, \ldots\}.$$

In other words, the one-to-one correspondence between X and $\mathbb{Z}^+$ enables us to use positive integers as "labels" for the elements of X.

A countably infinite set X is said to have the *cardinal number* $\aleph_0$ (*alaph nought*).

Definition 1.4.4. A set X, which is finite or countably infinite, is said to be *countable*. A set, which is not countable, is said to be *uncountable*.

Countable and uncountable sets are sometimes called *denumerable* and *non-denumerable sets*, respectively.

Theorem 1.4.1. *Every subset of a countable set is countable.*

Proof. Let X be the given countable set and $A \subseteq X$. If A is finite, we are through. So, let us suppose that A is infinite (so that X is also infinite). We can now write

$$X = \{x_1, x_2, x_3, \ldots\}.$$

Define a function k on $\mathbb{Z}^+$ as follows: Let $k(1)$ be the smallest positive integer such that $s_{k(1)} \in A$. Suppose the positive integers $k(1), k(2), \ldots, k(n-1)$ are already chosen. Let, now, $k(n)$ be the smallest integer $> k(n-1)$ such that $s_{k(n)} \in A$. Note that k is order-preserving, i.e., $k(m) > k(n)$ whenever $m > n$. Consider the composite function $s \circ k$. Its domain is $\mathbb{Z}^+$, and the range is A. Note that $s \circ k$ is one-to-one since

$$s[k(n)] = s[k(m)]$$

implies

$$s_{k(n)} = s_{k(m)},$$

which, in turn, implies

$$k(n) = k(m),$$

from which we conclude that

$$n = m$$

since k is order-preserving. This completes the proof.

We now give an example of an uncountable set.

Theorem 1.4.2. *The $\{\mathbb{R}\}$ of all real numbers is uncountable.*

Proof. It suffices to prove that $(0, 1)$, i.e., $\{x \in \mathbb{R} : 0 < x < 1\}$, is uncountable. If $(0, 1)$ were countable, then there exists a sequence $\{s_n\}$, whose terms would constitute the whole of $(0, 1)$, i.e., $(0, 1) = \{s_n\}$. We claim that this is not possible by exhibiting a real number in $(0, 1)$, which is not a term of the sequence $\{s_n\}$. Write each s_n as an infinite decimal:

$$s_n = 0.u_{n,1} u_{n,2} u_{n,3} \cdots,$$

where $u_{n,i} = 0, 1, 2, \ldots, 9$, $i = 1, 2, 3, \ldots$. Define the real number y by

$$y = 0.v_1 v_2 v_3 \cdots,$$

where

$$v_n = \begin{cases} 1, & \text{if } u_{n,n} \neq 1, \\ 2, & \text{if } u_{n,n} = 1. \end{cases}$$

Note that y differs from s_1 in the first decimal place, differs from s_2 in the second decimal place, $\ldots$, differs from s_n in the nth decimal place, $\ldots$. So, y cannot be a term of the sequence $\{s_n\}$, though $y \in (0, 1)$. This leads to a contradiction. Thus, $(0, 1)$ is uncountable. This completes the proof.

1.5 Set Algebras

Definition 1.5.1. Given the sets A_1, A_2, their *union*, denoted by $A_1 \cup A_2$, is defined as the set of all elements which belong to A_1 or A_2 or both.

In other words, $A_1 \cup A_2$ consists precisely of those elements which belong to at least one of A_1, A_2. Note that set union is commutative and associative, i.e.,

$$A_1 \bigcup A_2 = A_2 \bigcup A_1;$$

$$A_1 \bigcup (A_2 \bigcup A_3) = (A_1 \bigcup A_2) \bigcup A_3.$$

Now, we extend the above definition to any collection of sets.

Definition 1.5.2. If $\mathcal{F}$ is any collection of sets, the *union* of all sets in $\mathcal{F}$, denoted by

$$\bigcup_{A \in \mathcal{F}} A,$$

is defined as the set of all elements which belong to at least one of the sets in $\mathcal{F}$.

Definition 1.5.3. If $\mathcal{F}$ is any collection of sets, the *intersection* of all sets in $\mathcal{F}$, denoted by

$$\bigcap_{A \in \mathcal{F}} A,$$

is defined as the set of all elements which belong to every one of the sets in $\mathcal{F}$.

In particular, the intersection of two sets A_1, A_2, denoted by $A_1 \cap A_2$, consists of all elements common to both A_1 and A_2.

If $A_1 \cap A_2 = \emptyset$, then A_1 and A_2 are said to be *disjoint*. In this context, note that intersection of sets is commutative and associative, i.e.,

$$A_1 \bigcap A_2 = A_2 \bigcap A_1;$$

$$A_1 \bigcap (A_2 \bigcap A_3) = (A_1 \bigcap A_2) \bigcap A_3.$$

Definition 1.5.4. The *complement* of a set A relative to a set B, denoted by $B - A$, is defined as the set of all elements of B which do not belong to A, i.e.,

$$B - A = \{x : x \in B, x \notin A\}.$$

Theorem 1.5.1. *Let $\mathcal{F}$ be any collection of sets. Then, for every set B,*

$$B - \bigcup_{A \in \mathcal{F}} A = \bigcap_{A \in \mathcal{F}} (B - A)$$

and

$$B - \bigcap_{A \in \mathcal{F}} A = \bigcup_{A \in \mathcal{F}} (B - A).$$

Proof. Let $x \in B - \bigcup_{A \in \mathcal{F}} A$. Then, $x \in B$ and $x \notin \bigcup_{A \in \mathcal{F}} A$. So, $x \notin A$ for every $A \in \mathcal{F}$. Thus, $x \in (B - A)$ for every $A \in \mathcal{F}$. Consequently, $x \in \bigcap_{A \in \mathcal{F}} (B - A)$. Therefore, we have

$$B - \bigcup_{A \in \mathcal{F}} A = \bigcap_{A \in \mathcal{F}} (B - A). \tag{1.1}$$

Conversely, let $x \in \bigcap_{A \in \mathcal{F}} (B - A)$. So, $x \in B - A$ for every $A \in \mathcal{F}$, i.e., $x \in B$ and $x \notin A$ for every $A \in \mathcal{F}$, i.e., $x \in B$ and $x \notin \bigcup_{A \in \mathcal{F}} A$, i.e., $x \in B - \bigcup_{A \in \mathcal{F}} A$. Thus, we have

$$\bigcup_{A \in \mathcal{F}} (B - A) \subseteq B - \bigcup_{A \in \mathcal{F}} A. \tag{1.2}$$

Thus, (1.1) and (1.2) together imply that

$$B - \bigcup_{A \in \mathcal{F}} A = \bigcap_{A \in \mathcal{F}} (B - A).$$

The second statement is proved in a similar fashion. This completes the proof.

Definition 1.5.5. If $\mathcal{F}$ is a collection of sets such that every pair of distinct sets of $\mathcal{F}$ are disjoint, then $\mathcal{F}$ is called a *collection of disjoint sets*.

Theorem 1.5.2. *If $\mathcal{F}$ is a countable collection of disjoint sets, (say) $\mathcal{F} = \{A_1, A_2, A_3, \ldots\}$ such that each set A_n is countable, then $\bigcup_{k=1}^{\infty} A_k$ is also countable.*

Proof. Let

$$A_n = \{a_{1,n}, a_{2,n}, a_{3,n}, \ldots\} \text{ for every } n = 1, 2, \ldots$$

and $X = \bigcup_{k=1}^{\infty} A_k$. Now, every element $x \in X$ belongs to at least one A_k, and so

$$x = a_{m,n} \text{ for some pair of positive integers } (m, n).$$

Since the sets of $\mathcal{F}$ are pairwise disjoint, this pair of positive integers (m, n) is unique. Define the function

$$f : X \to \mathbb{Z}^+ \times \mathbb{Z}^+$$

by

$$f(x) = (m, n) \text{ for every } x = a_{m,n} \in X.$$

Note that f is well defined since (m, n) is uniquely determined by X. The range $f(X)$ of f is a subset of $\mathbb{Z}^+ \times \mathbb{Z}^+$, and so $f(X)$ is countable, since $\mathbb{Z}^+ \times \mathbb{Z}^+$ is countable. f is clearly (1-1), and so there is a one-to-one correspondence between the elements of X and $f(X)$. Since $f(X)$ is countable, it now follows that X is countable. This completes the proof.

Theorem 1.5.3. *Let* $\mathcal{F} = \{A_1, A_2, \ldots\}$ *be a countable collection of sets. Let* $G = \{B_1, B_2, \ldots\}$, *where*

$$B_1 = A_1, \quad B_n = A_n - \bigcup_{k=1}^{n-1} A_k \quad \text{for every } n = 2, 3, \ldots.$$

Then, $\mathscr{C}$ *is a countable collection of disjoint sets and*

$$\bigcup_{k=1}^{\infty} A_k = \bigcup_{k=1}^{\infty} B_k.$$

Proof. From the definition of B_n's, it follows that $\mathscr{C}$ is a collection of disjoint sets. Let $A = \bigcup_{k=1}^{\infty} A_k$ and $B = \bigcup_{k=1}^{\infty} B_k$. We claim that $A = B$. Let $x \in A$. Then, $x \in A_k$ for some k. Let n be the smallest positive integer such that $x \in A_n$ and $x \in \bigcup_{k=1}^{n-1} A_k$, i.e., $x \in A_n - \bigcup_{k=1}^{n-1} A_k$, $x \in B_n$, and so $x \in B$. Thus, we have

$$A \subseteq B.$$

Conversely, if $x \in B$, then $x \in B_n$ for some n, and so $x \in A_n$ for this n. Consequently, $x \in A$, which proves that

$$B \subseteq A.$$

So, $A = B$. This completes the proof.

As a consequence of Theorems 1.5.2 and 1.5.3, we have the following:

Theorem 1.5.4. *If* $\mathcal{F}$ *is a countable collection of countable sets, then the union of all sets in* $\mathcal{F}$ *is countable too, i.e., countable union of countable sets is countable.*

Theorem 1.5.5. *The set* Q *of all rational numbers is countable.*

Proof. Let A_n denote the set of all positive rational numbers with denominator n for every $n = 1, 2, \ldots$. Then, A_n is countable for every $n = 1, 2, \ldots$. Note that the set of all positive rational numbers is equal to $\bigcup_{n=1}^{\infty} A_n$. So, the set of all positive rational numbers is countable, in view of Theorem 1.5.4. It now follows that the set Q of all rational numbers is countable. This completes the proof.

Corollary 1.5.6. *The set of all irrational numbers is uncountable.*

1.6 Exercises

Excercise 1.1 Find which of the following relations is (a) reflexive, (b) symmetric, (c) transitive if

(1) $S = \mathbb{R} \times \mathbb{R}$ with $x < y$ for every $(x, y) \in S$;

(2) $S = \mathbb{R} \times \mathbb{R}$ with $x \leq y$ for every $(x, y) \in S$;

(3) $S = \mathbb{R} \times \mathbb{R}$ with $x < |y|$ for every $(x, y) \in S$;

(4) $S = \mathbb{R} \times \mathbb{R}$ such that $x^2 + y^2 = 1$;

(5) $S = \mathbb{R} \times \mathbb{R}$ such that $x^2 + x = y^2 = y$.

Excercise 1.2 Prove that equivalence among sets is an equivalence relation.

Excercise 1.3 F and G are defined for all real x by the given equation. In each case, when the composition function $G \circ F$ is defined, find the domain of $G \circ F$ and a formula for $(G \circ F)(x)$.

(1) $F(x) = 1 - x$, $G(x) = x^2 + 2x$;

(2) $F(x) = \begin{cases} 2x, & \text{if } 0 \leq x \leq 1, \\ 1, & \text{otherwise,} \end{cases}$ $G(x) = \begin{cases} x^2, & \text{if } 0 \leq x \leq 1, \\ 0, & \text{otherwise.} \end{cases}$

Excercise 1.4 Find $F(x)$ if

(1) $G(x) = x^3$ and $(G \circ F)(x) = x^3 - 3x^2 + 3x - 1$;

(2) $G(x) = x^2 + x + 3$ and $(G \circ F)(x) = x^2 - 3x + 5$.

Excercise 1.5 Prove the following identities for union and intersection of sets.

(1) (a) $(A \cup B) \cup C = A \cup (B \cup C)$;

    (b) $(A \cap B) \cap C = A \cap (B \cap C)$;

(2) $A \cap (B \cup C) = (A \cap B) \cup (A \cap C)$;

(3) $A \cup (B \cap C) = (A \cup B) \cap (A \cup C)$;

(4) $A \cap (B - C) = (A \cap B) - (A \cap C)$;

(5) $(A \cap B) - C = (A - C) \cap (B - C)$;

(6) $(A - B) \cup B = A$ if and only if $B \subseteq A$.

Excercise 1.6 Let $f : S \to T$ be a mapping. For every $A, B \subseteq S$, prove that

(1) $f(A \cup B) = f(A) \cup f(B)$;

(2) $f(A \cap B) \subseteq f(A) \cap f(B)$.

Excercise 1.7 Let $f : S \to T$ be a mapping. If $Y \subseteq T$, define

$$f^{-1}(Y) = \{x : x \in S, \ f(x) \in Y\}.$$

$f^{-1}(Y)$ is called the inverse image of Y under f. Prove the following:

(1) $X \subseteq f^{-1}[f(X)]$ for every $X \subseteq S$;

(2) $f[f^{-1}(Y)] \subseteq Y$ for every $Y \subseteq T$;

(3) $f^{-1}(Y_1 \cup Y_2) = f^{-1}(Y_1) \cup f^{-1}(Y_2)$ for every $Y_1, Y_2 \subseteq T$;

(4) $f^{-1}(Y_1 \cap Y_2) = f^{-1}(Y_1) \cap f^{-1}(Y_2)$ for every $Y_1, Y_2 \subseteq T$;

(5) $f^{-1}(T - Y) = S - f^{-1}(Y)$ for every $Y \subseteq T$.

Excercise 1.8 Let $f : S \to T$ be a mapping. Prove that the following statements are equivalent:

(1) f is (1-1) on S;

(2) $f(A \cap B) = f(A) \cap f(B)$ for every, $A, B \subseteq S$;

(3) $f^{-1}[f(A)] = A$ for every $A \subseteq S$;

(4) $f(A) \cap f(B) = \phi$ whenever $A \cap B = \phi$ for every $A, B \subseteq S$;

(5) $f(A - B) = f(A) - f(B)$ for every $A, B \subseteq S$ with $B \subseteq A$.

Excercise 1.9 If S is an infinite set, then prove that S contains a countably infinite subset.

Excercise 1.10 Prove that every infinite set S contains a proper subset equivalent to S.

Excercise 1.11 If A is countable and B is uncountable, then prove that $B - A$ is equivalent to B.

Excercise 1.12 Prove that $\mathbb{Z}^+ \times \mathbb{Z}^+$ is countable.

Excercise 1.13 Prove that the set of all irrational numbers is uncountable.

Excercise 1.14 Prove that the set of sequences of 0's and 1's is uncountable.

Excercise 1.15 Prove that the set of all circles in the plane, centers with rational Coordinates, and rational radii is countable.

Excercise 1.16 Prove that the set of all intervals with rational end points is countable.

2

The Real Number System

Mathematical analysis is concerned in some way to real numbers, and so we commence the study of analysis with a discussion of the real number system. In this study, we are not concerned about the construction of the real numbers. We are more concerned about the properties of the real numbers. So, we start by taking the real numbers for granted satisfying certain axioms from which further properties are derived. The construction of the real numbers system is an important part of the foundations of mathematics for which the reader can refer to Ref. [5].

Hereafter, we suppose that there exists a nonempty set $\mathbb{R}$ of elements, called *real numbers*, which satisfy ten axioms listed in the sequel. These axioms fall into three categories, viz. *field axioms*, *order axioms*, and the *completeness axiom* (also called *axiom of continuity* or *least upper bound axiom* (LUB)).

2.1   Field Axioms

Given $\mathbb{R}$, we suppose that there exist two operations, called *addition* and *multiplication*, such that for any $x, y \in \mathbb{R}$, the sum $x + y$ and product xy are real numbers uniquely determined by x, y which satisfy the following axioms, known as *field axioms*: For every $x, y, z \in \mathbb{R}$,

Axiom 1. $x + y = y + x$; $xy = yx$ (*commutative law*);

Axiom 2. $x + (y + z) = (x + y) + z$; $x(yz) = (xy)z$ (*associative law*);

Axiom 3. $x(y + z) = xy + xz$ (*distributive law*);

Axiom 4. Given $x, y \in \mathbb{R}$, there exists $z \in \mathbb{R}$ such that $x + z = y$. z is denoted by $y - x$. The number $x - x$ is denoted by 0. $0 - x$ is written as $-x$, and $-x$ is called the *negative* of x;

Axiom 5. There exists at least one $x \in \mathbb{R}$ such that $x \neq 0$. If $x, y \in \mathbb{R}$ with $x \neq 0$, there exists $z \in \mathbb{R}$ such that $xz = y$. z is denoted by $\frac{y}{x}$; the number $\frac{x}{x}$ is denoted by 1. If $x \neq 0$, $\frac{1}{x}$ is written as x^{-1}, called the *reciprocal* of x.

The usual laws of arithmetic can be worked out using the above five axioms: for instance, $x + (-y) = x - y$, $-(x - y) = y - x$, $-(-x) = x$, $(x^{-1})^{-1} = x$, and others.

2.2   Order Axioms

Given $\mathbb{R}$, we also suppose that there exists a relation, denoted by $<$, which establishes an ordering of real numbers and which satisfies the following axioms:

Axiom 6. For every $x, y \in \mathbb{R}$, exactly one of the relations $x < y$, $y < x$, $x = y$ (*law of trichotomy*) holds.
We also write $y > x$ for $x < y$.

Axiom 7. For every $x, y, z \in \mathbb{R}$, if $x < y$, then $x + z < y + z$ for every $z \in \mathbb{R}$;

Axiom 8. For every $x, y \in \mathbb{R}$, if $x, y > 0$, then $xy > 0$;

Axiom 9. For every $x, y, z \in \mathbb{R}$, if $x > y$ and $y > z$, then $x > z$ (*transitivity*).

A real number x is said to be *positive* if $x > 0$ and *negative* if $x < 0$. We shall denote the set of all positive real numbers by $\mathbb{R}^+$ and the set of all negative real numbers by $\mathbb{R}^-$.

Using the above axioms, we can derive the usual rules for inequalities; for example, if $x < y$ and $z > 0$, then $xz < yz$, while $xz > yz$ if $z < 0$; if $x > y$ and $z > w$, then $xz > yw$, if $y, z > 0$.

We write $x \le y$ to mean "$x < y$ or $x = y$". The symbol $\ge$ 1 is defined in a similar manner. A real number x is said to be *non-negative* if $x \ge 0$. The pair of inequalities $x < y$, $y < z$ is written as $x < y < z$.

The following result is extremely useful in analysis.

Theorem 2.2.1. *For every $a, b \in \mathbb{R}$ such that*

$$a \le b + \epsilon \quad \text{for every } \epsilon > 0, \tag{2.1}$$

we have

$$a \le b.$$

Proof. Suppose that $a \le b$ is not true. In view of Axiom 6, $b < a$. Let $\epsilon = \frac{a-b}{2}$. Then, $\epsilon > 0$. Now, we have

$$b + \epsilon = b + \frac{a-b}{2} = \frac{a+b}{2} < \frac{a+a}{2} = a \text{ since } b < a,$$

which contradicts (2.1). Thus, $a \le b$. This completes the proof.

2.3 Geometrical Representation of Real Numbers and Intervals

The real numbers are frequently represented geometrically as points on a line, called the *real line* or the *real axis*. A point is chosen to represent 0 and another to represent 1, as shown in the following figure:

This choice represents the scale. Under the set of axioms of Euclidean geometry, each point on the real line represents a unique real number and vice versa. We very often refer to the point x instead of the point representing the real number x. Now, the order relation has the following geometric interpretation. If $x < y$, x lies to the left of y on the real line.

Positive numbers lie to the right of 0 and negative numbers to the left of 0. If $a < b$, then x lies between a and b if and only if $a < x < b$.

The set of all points between a and b is called an *interval*.

Definition 2.3.1. Let $a < b$. Then

(1) The *open interval* (a, b) is defined by

$$(a, b) = \{x \in \mathbb{R} : a < x < b\};$$

(2) The *closed interval* $[a, b]$ is defined by

$$[a, b] = \{x \in \mathbb{R} : a \leq x \leq b\};$$

(3) The *half-open* and *half-closed* intervals $(a, b]$ and $[a, b)$ are defined as

$$(a, b] = \{x \in \mathbb{R} : a < x \leq b\};$$
$$[a, b) = \{x \in \mathbb{R} : a \leq x < b\};$$

(4) *The infinite intervals* are defined as follows:

$$(a, +\infty) = \{x \in \mathbb{R} : x > a\};$$
$$[a, +\infty) = \{x \in \mathbb{R} : x \geq a\};$$
$$(-\infty, a) = \{x \in \mathbb{R} : x < a\};$$
$$(-\infty, a] = \{x \in \mathbb{R} : x \leq a\}.$$

The real line $\mathbb{R}$ is sometimes called the *open interval* $(-\infty, +\infty)$. A single point is considered a degenerate closed interval. In the above context, note that the symbols $+\infty$ and $-\infty$ are used for convenience in notation and they are not considered as real numbers.

2.4   Integers, Rational Numbers, and Irrational Numbers

Definition 2.4.1. A set S of real numbers is called an *inductive set* if

(a) $1 \in S$;

and

(b) if $x \in S$, then $x + 1 \in S$.

Note that $\mathbb{R}, \mathbb{R}^+$ are inductive sets.

Definition 2.4.2. $n \in \mathbb{R}$ is called a *positive integer* if it belongs to every inductive set.

The set of all positive integers is denoted by $\mathbb{Z}^+$. We immediately note that $\mathbb{Z}^+$ is an inductive set. Also, $\mathbb{Z}^+$ is a subset of every inductive set, and so $\mathbb{Z}^+$ is the smallest inductive set. This property of $\mathbb{Z}^+$ is often known as the *principle of induction*. The negatives of positive integers are known as negative integers. The set of all positive integers, negative integers, and 0 is called the set of integers, denoted by $\mathbb{Z}$.

Quotients of integers of the form $\frac{a}{b}$, $b \neq 0$, are called *rational numbers*. The set of all rational numbers is denoted by $\mathbb{Q}$. Note that $\mathbb{Z} \subsetneq \mathbb{Q}$. Real numbers, which are not rational, are called *irrational numbers*.

Examples of irrational numbers are $\sqrt{2}$, e, π, e^π (we do not go into the details of proving that these numbers are induced irrational). $\mathbb{Q}$ together with irrational numbers make up the whole of $\mathbb{R}$.

2.5   Upper Bounds, Least Upper Bound or Supremum, the Completeness Axiom, Archimedean Property of $\mathbb{R}$

Definition 2.5.1. Let S be a set of real numbers. If there exists $b \in \mathbb{R}$ such that

$$x \leq b \text{ for every } x \in S,$$

then b is called an *upper bound* of S, and in this case, we say that S is *bounded above* by b.

We note that if b is an upper bound of S, then any $b' \in \mathbb{R}$ such that $b' > b$ is also an upper bound of S. If b is an upper bound of S and $b \in S$, then b is called the *largest element* or *maximum element* of S. When such a maximum element b exists, it is unique, written as

$$b = \max \ S.$$

A set with no upper bound is said to be *unbounded above*.

The terms lower bound, bounded below, smallest or minimum element are defined in a similar fashion. If S has a minimum element, it is unique, denoted by min S. S is said to be bounded if S is bounded above and bounded below. Equivalently, $S \subseteq [-a, a]$ for some $a > 0$.

Definition 2.5.2. Let S be a set of real numbers bounded above. $b \in \mathbb{R}$ is called a *least upper bound* of S or *lub* of S if

(a) b is an upper bound of S;

and

(b) no number less than b is an upper bound of S.

Note that, whenever least upper bound of S exists, it is unique and so we speak of the least upper bound of S. The least upper bound of S is also called the *supremum* of S, written as sup S. If S has a maximum element, then

$$\sup\ S = \max\ S.$$

The *greatest lower bound* or *glb* or *infimum* of S is defined in a similar manner. The infimum of S is written as inf S whenever it exists.

The final axiom of the real number system involves the notion of supremum.

Axiom 10. (*Completeness Axiom*) Every non-empty set S of real numbers which is bounded above has a supremum, i.e., there exists $b \in \mathbb{R}$ such that

$$b = \sup\ S.$$

As a consequence of Axiom 10, it now follows that any non-empty set S of real numbers which is bounded below has an infimum.

As a consequence of Axiom 10, we also have the following important result.

Theorem 2.5.1. *The set $\mathbb{Z}^+$ of all positive integers is unbounded above.*

Proof. Suppose that $\mathbb{Z}^+$ is bounded above. Using Axiom 10, $\mathbb{Z}^+$ has a supremum. Let $b = \sup\ \mathbb{Z}^+$. In view of Theorem 2.2.1, $b - 1$ cannot be an upper bound of $\mathbb{Z}^+$. So, there exists $n \in \mathbb{Z}^+$ such that

$$n > b - 1, \quad \text{i.e.,} \quad n + 1 > b,$$

where $n + 1 \in \mathbb{Z}^+$, which contradicts the fact that $b = \sup\ \mathbb{Z}^+$, completing the proof.

Corollary 2.5.2. *For every $x \in \mathbb{R}$, there exists $n \in \mathbb{Z}^+$ such that*

$$n > x.$$

Theorem 2.5.3. [The Archimedean Property of $\mathbb{R}$] *If $x, y \in \mathbb{R}$ with $x > 0$, then there exists $n \in \mathbb{Z}^+$ such that*

$$nx > y.$$

Proof. In Corollary 2.5.2, replace x by $\frac{y}{x}$. Then there exists $n \in \mathbb{Z}^+$ such that

$$n > \frac{y}{x}, \quad \text{i.e.,} \quad nx > y.$$

Geometrically, Theorem 2.5.3 means that any line segment on the real line, however long, can be covered by a finite number of line segments of a given positive length, no matter how small. This is usually referred to the Archimedean property of the real number system.

2.6 Infinite Decimal Representation of Real Numbers

The following result shows that real numbers can be approximated to any degree of accuracy by rational numbers with finite decimal expansion.

Theorem 2.6.1. *Let $x \geq 0$. Then, for every $n = 1, 2, \ldots$, there is a finite decimal*

$$r_n = a_0.a_1a_2 \cdots a_n$$

such that

$$r_n \leq x < r_n + \frac{1}{10^n}.$$

Proof. Let $x \geq 0$ and S be the set of all non-negative integers $\leq x$. Then, S is non-empty with $0 \in S$. Also, S is finite. Let $a_0 = \max S$. Note that $a_0 \in S$ and a_0 is a non-negative integer. We call a_0 the greatest integer $\leq x$ and write

$$a_0 = [x].$$

We also call a_0 the integral part of x. We observe that

$$a_0 \leq x < a_0 + 1.$$

Let

$$a_1 = [10x - 10a_0].$$

Now, we have

$$0 \leq 10x - 10a_0 = 10(x - a_0) < 10,$$

so that

$$0 \leq a_1 \leq 9$$

and
$$a_1 \leq 10x - 10a_0 < a_1 + 1,$$

i.e., a_1 is the greatest integer satisfying the inequalities

$$a_0 + \frac{a_1}{10} \leq x < a_0 + \frac{a_1 + 1}{10}.$$

Having chosen a_i, $0 \leq a_i \leq 9$, $i = 1, 2, \ldots, (n-1)$, let a_n be the greatest integer satisfying the inequalities

$$a_0 + \frac{a_1}{10} + \cdots + \frac{a_n}{10^n} \leq x < a_0 + \frac{a_1}{10} + \cdots + \frac{a_n + 1}{10^n}. \qquad (2.2)$$

Then, we have
$$0 \leq a_n \leq 9$$

and
$$r_n \leq x < r_n + \frac{1}{10^n},$$

where

$$r_n = a_0 + \frac{a_1}{10} + \cdots + \frac{a_n}{10^n}$$
$$= a_0.a_1 a_2 \cdots a_n.$$

This completes the proof.

Incidentally, we note that x is the supremum of the rational numbers $r_1, r_2, \ldots$.

The integers $a_0, a_1, a_2, \ldots$, obtained in Theorem 2.6.1 help us to define an infinite decimal expansion of x, i.e., we can now write

$$x = a_0.a_1 a_2 \cdots,$$

where a_n is the greatest integer satisfying the inequalities (2.2) for every $n = 1, 2, \ldots$. For instance, we have

$$\frac{1}{16} = 0.0625000 \cdots$$

with $a_0 = a_1 = 0$, $a_2 = 6$, $a_3 = 2$, $a_4 = 5$, $a_n = 0$, $n \geq 5$. If we interchange the inequalities $\leq$ and $<$ in (2.2), we get a different definition of decimal expansions. The rational numbers r_n satisfy the inequalities

$$r_n < x \leq r_n + \frac{1}{10^n} \quad \text{for every} \quad n = 1, 2, \ldots.$$

The integers $a_0, a_1, a_2, \ldots$ may not be the same as those in (2.2). For instance, if we apply this new definition to $\frac{1}{16}$, we obtain

$$\frac{1}{16} = 0.0624999 \cdots.$$

Incidentally, we note that

$$0.0624999\cdots = \frac{6}{10^2} + \frac{2}{10^3} + \frac{4}{10^4}$$
$$+ \frac{9}{10^5} + \frac{9}{10^6} + \frac{9}{10^7} + \cdots$$
$$= \frac{6}{100} + \frac{2}{1000} + \frac{4}{10000}$$
$$+ \frac{9}{10^5}\left[1 + \frac{1}{10} + \frac{1}{10^2} + \cdots\right]$$
$$= \frac{600 + 20 + 4}{10^4} + \frac{9}{10^5}\frac{1}{\left(1 - \frac{1}{10}\right)}$$
$$= \frac{624}{10^4} + \frac{9}{10^5}\frac{10}{9}$$
$$= \frac{624}{10^4} + \frac{1}{10^4}$$
$$= \frac{625}{10^4}$$
$$= \frac{625}{10000}$$
$$= \frac{1}{16}.$$

We do not distinguish between the two expansions of $\frac{1}{16}$. We note that a real number may have two different decimal expansions, since two different sets of real numbers $r_1, r_2, \ldots$ have the same supremum.

2.7 Absolute Value, Triangle Inequality, Cauchy–Schwarz Inequality

Definition 2.7.1. If $x \in \mathbb{R}$, then the *absolute value* of x, denoted by $|x|$, is defined by

$$|x| = \begin{cases} x, & \text{if } x \geq 0, \\ -x, & \text{if } x < 0. \end{cases}$$

Theorem 2.7.1. *Let $a \geq 0$. Then, $|x| \leq a$ if and only if $-a \leq x \leq a$.*

Proof. In view of Definition 2.7.1, we have

$$-|x| \leq x \leq |x|.$$

Let, now, $|x| \leq a$. So, we have

$$-a \leq -|x| \leq x \leq |x| \leq a.$$

Thus, it follows that

$$-a \leq x \leq a.$$

Conversely, let $-a \leq x \leq a$. If $x \geq 0$, then $|x| = x \leq a$, while if $x < 0$, then $|x| = -x \leq a$, so that $|x| \leq a$ in all cases. This completes the proof.

Theorem 2.7.2. (The Triangle Inequality) *If $x, y \in \mathbb{R}$, then we have*

$$|x + y| \leq |x| + |y|.$$

Proof. Now, we have

$$-|x| \leq x \leq |x|$$

and

$$-|y| \leq y \leq |y|.$$

Adding the inequalities above, we have

$$-(|x + y|) \leq x + y \leq |x| + |y|.$$

In view of Theorem 2.7.1, it follows that

$$|x + y| \leq |x| + |y|.$$

This completes the proof.

Corollary 2.7.3. *For every $x, y \in \mathbb{R}$, we have*

$$|x + y| \geq ||x| - |y||.$$

Proof. By using Theorem 2.7.2, we have

$$|x| = |(x - y) + y| \leq |x - y| + |y|,$$

which implies

$$|x| - |y| \leq |x - y|.$$

Interchanging the roles of x, y, we have

$$|y| - |x| \leq |y - x| = |x - y|$$

and so

$$(|x| - |y|) \leq |x - y|, \quad \text{i.e.,} \quad |x| - |y| \leq |x - y|.$$

Thus, we have

$$-|x - y| \leq |x| - |y| \leq |x - y|.$$

In view of Theorem 2.7.1, it follows that

$$||x| - |y|| \leq |x - y|.$$

This completes the proof.

Remark 2.7.1. Interpreting geometrically, Theorem 2.7.2 means that, in any triangle, the length of any side is less than or equal to the sum of the lengths of the other two sides. Corollary 2.7.3 means that, in any triangle, the length of any side is greater than or equal to the difference in the lengths of the other two sides.

By induction, we have

$$|x_1 + x_2 + \cdots + x_n| \le |x_1| + |x_2| + \cdots + |x_n|$$

and

$$|x_1 + x_2 + \cdots + x_n| \ge |x_1| - |x_2| - \cdots - |x_n|$$

for every $x_i \in \mathbb{R}$, $i = 1, 2, \ldots, n$.

The following inequality is very useful in analysis:

Theorem 2.7.4. (The Cauchy–Schwarz Inequality) *If $a_k, b_k \in \mathbb{R}$ for every $k = 1, 2, \ldots, n$, then we have*

$$\left(\sum_{k=1}^{n} a_k b_k \right)^2 \le \left(\sum_{k=1}^{n} a_k^2 \right) \left(\sum_{k=1}^{n} b_k^2 \right).$$

Proof. First of all, we note that

$$\sum_{k=1}^{n} (a_k x + b_k)^2 \ge 0 \quad \text{for every } \ x \in \mathbb{R}.$$

Expanding the left-hand side, we get

$$Ax^2 + 2Bx + C \ge 0,$$

where

$$A = \sum_{k=1}^{n} a_k^2, \quad B = \sum_{k=1}^{n} a_k b_k, \quad C = \sum_{k=1}^{n} b_k^2.$$

If $A = 0$, then $a_k = 0$ for every $k = 1, 2, \ldots, n$, so that the inequality holds trivially.

If $A > 0$, then put $x = -\frac{B}{A}$. We obtain

$$B^2 \le AC,$$

i.e.,

$$\left(\sum_{k=1}^{n} a_k b_k \right)^2 \le \left(\sum_{k=1}^{n} a_k^2 \right) \left(\sum_{k=1}^{n} b_k^2 \right).$$

This completes the proof.

2.8   Extended Real Number System $\mathbb{R}^*$

We now extend the real number system $\mathbb{R}$ by adjoining two *"ideal points"*, denoted by $+\infty$ and $-\infty$, called *"plus infinity"* and *"minus infinite"*, respectively.

Definition 2.8.1. By the extended real number system $\mathbb{R}^*$, we mean the set of all real numbers $\mathbb{R}$ together with the two symbols $+\infty$ and $-\infty$ satisfying the following properties:

(1) If $x \in \mathbb{R}$, then we have

$$x + (+\infty) = +\infty;$$
$$x + (-\infty) = -\infty;$$
$$x - (+\infty) = -\infty;$$
$$x - (-\infty) = +\infty;$$
$$\frac{x}{+\infty} = \frac{x}{-\infty} = 0;$$

(2) If $x > 0$, then we have

$$x(+\infty) = +\infty;$$
$$x(-\infty) = -\infty;$$

(3) If $x < 0$, then we have

$$x(+\infty) = -\infty;$$
$$x(-\infty) = +\infty;$$

(4) If $x \in \mathbb{R}$, then we have

$$-\infty < x < +\infty;$$

(5) Also, we have

$$(+\infty) + (+\infty) = (+\infty)(+\infty) = (-\infty)(-\infty) = +\infty;$$
$$(-\infty) + (-\infty) = (+\infty)(-\infty) = -\infty.$$

Sometimes, we denote $\mathbb{R}$ by $(-\infty, +\infty)$ and $\mathbb{R}^*$ by $[+\infty, +\infty]$.

We note that the reason for introducing the symbols $+\infty$ and $-\infty$ is a matter of convenience. For instance, if we define $+\infty$ to be the supremum of a set of real numbers not bounded above, then every non-empty subset of $\mathbb{R}$ has a supremum in $\mathbb{R}^*$. The supremum is finite or $+\infty$ according as the set is bounded above or not. In a similar manner, we define $-\infty$ as the infimum of any set of real numbers not bounded below, so that every non-empty subset of $\mathbb{R}$ has an infimum in $\mathbb{R}^*$.

2.9   Exercises

Excercise 2.1 Find max S, min S, sup S, inf S if

(1) $S = [0, 1]$;

(2) $S = [0, 1)$;

(3) $S = (0, 1]$;

(4) $S = (0, 1)$.

Excercise 2.2 Let A, B be non-empty subsets of $\mathbb{R}$. Let

$$C = \{x + y : x \in A,\ y \in B\}.$$

If A, B have supremum, prove that C has supremum and

$$\sup C = \sup A + \sup B \quad (\textit{additive property}).$$

Excercise 2.3 Let S, T be non-empty subsets of $\mathbb{R}$ such that

$$s \leq t \ \text{ for every } s \in S \text{ and } t \in T.$$

If T has a supremum, prove that S has a supremum and

$$\sup S \leq \sup T \quad (\textit{comparison property}).$$

Excercise 2.4 Prove that any non-empty set of positive integers contains a smallest element (*Well-ordering Principle*).

Excercise 2.5 Prove that $\sqrt{2}$, $\sqrt{3}$, $\sqrt{2} + \sqrt{3}$ are irrational.

Excercise 2.6 Find the rational number, whose decimal expansion is

$$0.334444\cdots$$

Excercise 2.7 Given any real number $x > 0$, prove that there is an irrational number between 0 and x.

Excercise 2.8 Prove that, for any integer $n \geq 1$, $\sqrt{n-1} + \sqrt{n+1}$ is irrational.

Excercise 2.9 Prove that the supremum and infimum of a set of real numbers are unique, whenever they exist.

Excercise 2.10 Let A, B be two sets of positive numbers which are bounded above. Let $a = \sup A$ and $b = \sup B$. Let

$$C = \{xy : x \in A,\ y \in B\}.$$

Prove that C is bounded above and $\sup C = ab$.

3

Sequences and Series of Real Numbers

In this chapter, we give some convergence theorems of the sequences and series of real numbers, some convergence tests of series, the rearrangements, and Cauchy multiplications of series.

3.1 Convergent and Divergent Sequences of Real Numbers

Definition 3.1.1. A sequence $\{x_n\}$ of real numbers is said to *converge* if there exists $x \in \mathbb{R}$ such that, for every $\epsilon > 0$, there exists a positive integer N, depending on ϵ, such that

$$|x_n - x| < \epsilon \text{ whenever } n \geq N.$$

In such a case, we say that $\{x_n\}$ *converges* to x, written as

$$\lim_{n \to \infty} x_n = x.$$

Note that x, whenever it exists, is unique. It is called the *limit* of $\{x_n\}$. The sequence $\{x_n\}$ is said to *diverge* if it is not convergent.

Definition 3.1.2. A sequence $\{x_n\}$ is called a *Cauchy sequence* if, for every $\epsilon > 0$, there exists a positive integer N such that

$$|x_m - x_n| < \epsilon \text{ whenever } m, n \geq N.$$

We will see in Chapter 4 that $\mathbb{R}$ is a complete metric space and so a sequence in $\mathbb{R}$ converges if and only if it is a Cauchy sequence.

Note that whenever the limit to which a sequence $\{x_n\}$ converges is not known, it is very useful to prove that $\{x_n\}$ is a Cauchy sequence and immediately conclude that $\{x_n\}$ converges.

Definition 3.1.3. A sequence $\{x_n\}$ is said to be *bounded* if there exists $M > 0$ such that

$$|x_n| \leq M \text{ for every } n = 1, 2, \ldots.$$

Note that $\{x_n\}$ is bounded if and only if $\{x_n\}$ is both bounded above and bounded below.

Theorem 3.1.1. *A convergent sequence $\{x_n\}$ is bounded.*

Proof. Suppose that $\{x_n\}$ converges to x (say). Then, there exists a positive integer N such that

$$|x_n - x| < 1 \quad \text{for every } n \geq N.$$

So, it follows that, for every $n \geq N$,

$$\begin{aligned}
|x_n| &= |(x_n - x) + x| \\
&\leq |x_n - x| + |x| \\
&< 1 + |x|.
\end{aligned}$$

Let

$$M = \max(|x_1|, |x_2|, \ldots, |x_{N-1}|, 1 + |x|).$$

Then, we have

$$|x_n| \leq M \quad \text{for every } n = 1, 2, \ldots,$$

i.e., $\{x_n\}$ is bounded. This completes the proof.

Note that the converse is not true. For instance, the sequence $\{1, 0, 1, 0, \ldots\}$ is bounded, but it is not convergent. It follows from Theorem 3.1.1 that an unbounded sequence diverges. One can easily prove that if $\{x_n\}$ converges to x, then every subsequence $\{x_{k(n)}\}$ of $\{x_n\}$ also converges to x. So, if two subsequences of a sequence $\{x_n\}$ converge to two different limits, then the sequence necessarily diverges.

Definition 3.1.4. A sequence $\{x_n\}$ of real numbers is said to *diverge* to $+\infty$ if, for every $M > 0$, there exists a positive integer N, depending on M, such that

$$x_n > M \quad \text{whenever } n \geq N.$$

In this case, we write

$$\lim_{n \to \infty} x_n = +\infty.$$

If $\lim\limits_{n \to \infty} (-x_n) = +\infty$, we write $\lim\limits_{n \to \infty} x_n = -\infty$ and we say that $\{x_n\}$ diverges to $-\infty$. Note that there are divergent sequences of real numbers, which do not diverge to $+\infty$ or $-\infty$. For instance, the divergent sequence $\{1, 0, 1, 0, \ldots\}$ does not diverge to $+\infty$ or $-\infty$.

3.2 Limit Superior and Limit Inferior of a Sequence of Real Numbers

Definition 3.2.1. Let $\{a_n\}$ be a sequence of real numbers. Then, a real number U is called the *limit superior* or the *upper limit* of $\{a_n\}$ if U satisfies the following conditions:

(a) for every $\epsilon > 0$, there exists a positive integer N such that

$$a_n < U + \epsilon \quad \text{for every } n > N;$$

(b) there are an infinity of n's such that

$$a_n > U - \epsilon.$$

We write

$$U = \limsup_{n \to \infty} a_n = \overline{\lim_{n \to \infty}} a_n.$$

The statement (a) implies that $\{a_n\}$ is bounded above. If $\{a_n\}$ is not bounded above, then we define

$$\limsup_{n \to \infty} a_n = +\infty.$$

If $\{a_n\}$ is bounded above, but not bounded below, and $\{a_n\}$ has no finite limit superior, then we define

$$\underline{\lim_{n \to \infty}} a_n = -\infty.$$

The *limit inferior* or the *lower limit* of $\{a_n\}$ is defined in a similar way.

In Definition 3.2.1, the statement (a) means that ultimately all the terms of the sequence $\{a_n\}$ lie to the left of $U + \epsilon$, while the statement (b) means that infinitely many terms of $\{a_n\}$ lie to the right of $U - \epsilon$. Note that, whenever the limit superior of $\{a_n\}$ and limit inferior of $\{a_n\}$ exist, they are unique.

Definition 3.2.2. A sequence $\{a_n\}$ is said to *oscillate* if

$$\limsup_{n \to \infty} a_n \neq \liminf_{n \to \infty} a_n.$$

Theorem 3.2.1. *For any sequence* $\{a_n\}$ *of real numbers, we have*

$$\liminf_{n \to \infty} a_n \neq \limsup_{n \to \infty} a_n.$$

Proof. We consider the case when both $\liminf_{n \to \infty} a_n$ and $\limsup_{n \to \infty} a_n$ exist finitely, leaving the other cases to the reader. Let

$$\liminf_{n \to \infty} a_n = L \quad \text{and} \quad \limsup_{n \to \infty} a_n = U.$$

Then, we suppose that $L, U < \infty$. From Definition 3.2.1, given $\epsilon > 0$, there exists a positive integer N_1 such that

$$a_n < U + \frac{\epsilon}{2} \quad \text{for every } n > N_1.$$

Similarly, there exists a positive integer N_2 such that

$$a_n > L - \frac{\epsilon}{2} \quad \text{for every } n > N_2.$$

Let $N = \max\{N_1, N_2\}$. Then, for every $n > N$, we have

$$L - \frac{\epsilon}{2} < a_n < U + \frac{\epsilon}{2},$$

which implies

$$L < U + \epsilon \quad \text{for every} \quad \epsilon > 0.$$

Now, appealing to Theorem 2.3.1, we have

$$L \leq U.$$

This completes the proof.

The following result can be easily proved and left as an exercise to the reader:

Theorem 3.2.2. *The sequence $\{a_n\}$ of real numbers converges if and only if* $\liminf\limits_{n\to\infty} a_n$, $\limsup\limits_{n\to\infty} a_n$ *exist finitely and are equal. In this case, we have*

$$\liminf_{n\to\infty} a_n = \limsup_{n\to\infty} a_n = \lim_{n\to\infty} a_n.$$

3.3   Infinite Series of Real Numbers

Definition 3.3.1. Let $\{a_n\}$ be a given sequence of real numbers. Define a new sequence $\{s_n\}$ by

$$s_n = a_1 + a_2 + \cdots + a_n = \sum_{k=1}^{n} a_k \quad \text{for every } n = 1, 2, \ldots.$$

(1) The ordered pair of sequences $(\{a_n\}, \{s_n\})$ is called an *infinite series*;

(2) The term s_n is called the *nth partial sum* of the series;

(3) The sequence $\{s_n\}$ is called the *sequence of partial sums* associated with the series;

(4) The series is said to *converge* or *diverge* according as $\{s_n\}$ converges or diverges;

(5) The series is denoted by

$$a_1 + a_2 + \cdots + a_n + \cdots, \quad a_1 + a_2 + a_3 + \cdots, \quad \sum_{k=1}^{n} a_k;$$

(6) If the sequence $\{s_n\}$ of partial sums converges to s, s is called the *sum* of the series, written as

$$\sum_{k=1}^{n} a_k = s.$$

The following definition is useful:

Definition 3.3.2. Let $\{a_n\}$ be a sequence of real numbers.

(1) The sequence $\{a_n\}$ is said to be *increasing* or *non-decreasing* if

$$a_n \leq a_{n+1} \text{ for every } n = 1, 2, \ldots,$$

in which case, we write $a_n \nearrow$;

(2) The sequence $\{a_n\}$ is said to be *decreasing* or *non-increasing* if

$$a_n \geq a_{n+1} \text{ for every } n = 1, 2, \ldots,$$

in which case, we write $a_n \searrow$;

(3) The sequence $\{a_n\}$ is said to be *monotonic* if it is increasing or decreasing;

(4) We also say that $\{a_n\}$ is *strictly increasing* if

$$a_n < a_{n+1} \text{ for every } n = 1, 2, \ldots$$

and *strictly decreasing* if

$$a_n > a_{n+1} \text{ for every } n = 1, 2, \ldots.$$

Theorem 3.3.1. *Let $a_n \geq 0$ for every $n = 1, 2, \ldots$. Then, the following are equivalent:*

(1) $\displaystyle\sum_{k=1}^{\infty} a_k$ *converges;*

(2) *The sequence $\{s_n\}$ of partial sums is bounded above.*

Proof. $(1) \implies (2)$ Let $s_n = \sum_{k=1}^{n} a_k$ for every $n = 1, 2, \ldots$. Since $a_n \geq 0$, we have

$$s_{n+1} = s_n + a_{n+1} \geq s_n \text{ for every } n = 1, 2, \ldots,$$

so that $\{s_n\}$ is non-decreasing. If $\sum_{k=1}^{\infty} a_k$ converges, then $\{s_n\}$ converges, and so $\{s_n\}$ is bounded above by using Theorem 3.1.1.

$(2) \implies (1)$ Let $\{s_n\}$ be bounded above and let

$$\sup s_n = U.$$

Then, we have

$$s_n \leq U \quad \text{for every } n = 1, 2, \ldots.$$

Given $\epsilon > 0$, $U - \epsilon$ cannot be an upper bound of $\{a_n\}$. So, we have

$$s_N > U - \epsilon \quad \text{for some positive integer } N.$$

Since $\{s_n\}$ is non-decreasing, it follows that, for every $n \geq N$,

$$s_n \geq s_N > U - \epsilon.$$

So, we have

$$U - \epsilon < s_n \leq U < U + \epsilon \quad \text{for every } n \geq N,$$

i.e.,

$$|s_n - U| < \epsilon \quad \text{for every } n \geq N,$$

which implies

$$\lim_{n \to \infty} s_n = U.$$

Therefore, we have

$$\sum_{k=1}^{\infty} a_k = U.$$

This completes the proof.

Theorem 3.3.2. (The Cauchy Condition for Series) *The following are equivalent:*

(1) *The series $\sum_{n=1}^{\infty} a_n$ converges;*

(2) *For every $\epsilon > 0$, there exists a positive integer N such that, for every $n > N$,*

$$|a_{n+1} + a_{n+2} + \cdots + a_{n+p}| < \epsilon \quad \text{for every } p = 1, 2, \ldots. \qquad (3.1)$$

Proof. (2) $\implies$ (1) Let $s_n = \sum_{k=1}^{n} a_k$ for every $n = 1, 2, \ldots$. Let $\sum_{k=1}^{\infty} a_k$ converge, i.e., $\{s_n\}$ converges, and so $\{s_n\}$ is a Cauchy sequence. Thus, given $\epsilon > 0$, there exists a positive integer N such that, for every $n > N$,

$$|s_{n+p} - s_n| < \epsilon \quad \text{for every } p = 1, 2, \ldots,$$

i.e.,

$$|a_{n+1} + a_{n+2} + \cdots + a_{n+p}| < \epsilon \quad \text{for every } p = 1, 2, \ldots.$$

So, (3.1) holds.

(1) $\implies$ (2) Let (3.1) hold. Then, we have

$$|s_{n+p} - s_n| < \epsilon \quad \text{for every } n > N, \, p = 1, 2, \ldots.$$

It now follows that $\{s_n\}$ is a Cauchy sequence of real numbers. In Chapter 4, it will be proved that $\mathbb{R}$ is complete, i.e., every Cauchy sequence in $\mathbb{R}$ converges to a real number. So, $\{s_n\}$ converges, i.e., $\sum_{k=1}^{\infty} a_k$ converges. This completes the proof.

Corollary 3.3.3. *Suppose that a series $\sum_{k=1}^{\infty} a_k$ converges. Then, $\{s_n\}$ converges, where $s_n = \sum_{k=1}^{n} a_k$ for every $n = 1, 2, \ldots$, and so $\{s_n\}$ is a Cauchy sequence.*

Thus, we have

$$s_{n-1} \to 0 \quad \text{as } n \to \infty,$$

which implies

$$a_n \to 0 \quad \text{as } n \to \infty,$$

i.e., $\lim_{n \to \infty} a_n = 0$ is a necessary condition for the convergence of the series $\sum_{n=1}^{\infty} a_n$. The converse is not true, i.e., the above condition is not sufficient. For instance, consider the series $\sum_{n=1}^{\infty} a_n$, where

$$a_n = \frac{1}{n} \quad \text{for every } n = 1, 2, \ldots.$$

When $n = p = 2^m$ in (3.1), we find

$$a_{n+1} + a_{n+2} + \cdots + a_{n+p} = \frac{1}{2^m + 1} + \frac{1}{2^m + 2} + \cdots + \frac{1}{2^m + 2^m},$$

$$\frac{2^m}{2^m + 2^m} = \frac{1}{2} \quad \text{for every } m = 1, 2, \ldots.$$

Consequently, the Cauchy condition (3.1) is not satisfied for $\epsilon \leq \frac{1}{2}$. Thus, $\sum_{n=1}^{\infty} \frac{1}{n}$ diverges. Note that $\lim_{n \to \infty} \frac{1}{n} = 0$. The series $\sum_{n=1}^{\infty} \frac{1}{n}$ is called the "*harmonic series*".

Definition 3.3.3. Let $\sum_{n=1}^{\infty} a_n$ be a series of real numbers.

(1) A series $\sum_{n=1}^{\infty} a_n$ of real numbers is said to be *absolutely convergent* if $\sum_{n=1}^{\infty} |a_n|$ converges;

(2) The series $\sum_{n=1}^{\infty} a_n$ is said to be *conditionally convergent* if $\sum_{n=1}^{\infty} a_n$ converges and $\sum_{n=1}^{\infty} |a_n|$ diverges.

Theorem 3.3.4. *The absolute convergence of $\sum_{n=1}^{\infty} a_n$ implies the convergence of $\sum_{n=1}^{\infty} a_n$.*

Proof. Note that

$$|a_{n+1} + a_{n+2} + \cdots + a_{n+p}| \leq |a_{n+1}| + |a_{n+2}| + \cdots + |a_{n+p}|$$

and then appeal to Theorem 3.3.2. This completes the proof.

Note that the converse of Theorem 3.3.4 is not true. Consider the series $\sum_{n=1}^{\infty} \frac{(-1)^{n+1}}{n}$, i.e., $1 - \frac{1}{2} + \frac{1}{3} - \frac{1}{4} + \cdots$. Now, we prove that it converges. It does not converge absolutely since $\sum_{n=1}^{\infty} \frac{(-1)^{n+1}}{n} = \sum_{n=1}^{\infty} \frac{1}{n}$, which diverges (proved earlier). We also note that $\sum_{n=1}^{\infty} \frac{(-1)^{n+1}}{n}$ is an example of a conditionally convergent series. To prove that $\sum_{n=1}^{\infty} \frac{(-1)^{n+1}}{n}$ converges, we proceed as follows:

We claim that

$$0 < \frac{1}{n+1} - \frac{1}{n+2} + \frac{1}{n+3} - \frac{1}{n+4} + \cdots + \frac{(-1)^{p+1}}{n+p} < \frac{1}{n+1}. \qquad (3.2)$$

Let p be even (say) $= 2k$. Then, we have

$$\frac{1}{n+1} - \frac{1}{n+2} + \frac{1}{n+3} - \frac{1}{n+4} + \cdots + \frac{(-1)^{p+1}}{n+p}$$
$$= \frac{1}{n+1} - \frac{1}{n+2} + \frac{1}{n+3} - \frac{1}{n+4} + \cdots + \frac{1}{n+p-1} - \frac{1}{n+p}$$
$$= \left(\frac{1}{n+1} - \frac{1}{n+2}\right) + \left(\frac{1}{n+3} - \frac{1}{n+4}\right) + \cdots + \left(\frac{1}{n+p-1} - \frac{1}{n+p}\right)$$
$$= \frac{1}{(n+1)(n+2)} + \frac{1}{(n+3)(n+4)} + \cdots + \frac{1}{(n+p-1)(n+p)}$$
$$> 0,$$

where there are k pairs of terms. Again, in this case, we have

$$-\frac{1}{n+2} + \frac{1}{n+3} - \frac{1}{n+4} + \frac{1}{n+5} + \cdots + \frac{(-1)^{p+1}}{n+p}$$
$$= -\frac{1}{n+2} + \frac{1}{n+3} - \frac{1}{n+4} + \frac{1}{n+5}$$
$$\quad - \frac{1}{n+p-2} + \frac{1}{n+p-1} - \frac{1}{n+p}$$
$$= -\left(\frac{1}{n+2} - \frac{1}{n+3}\right) - \left(\frac{1}{n+4} - \frac{1}{n+5}\right)$$
$$\quad - \cdots - \left(\frac{1}{n+p-2} - \frac{1}{n+p-1}\right) - \frac{1}{n+p}$$
$$= -\left[\frac{1}{(n+2)(n+3)} + \frac{1}{(n+4)(n+5)}\right.$$
$$\quad \left. + \cdots + \frac{1}{(n+p-2)(n+p-1)} + \frac{1}{n+p}\right]$$
$$< 0$$

and so

$$\frac{1}{n+1} - \frac{1}{n+2} + \frac{1}{n+3} - \frac{1}{n+4} + \cdots + \frac{(-1)^{p+1}}{n+p} < \frac{1}{n+1}.$$

Thus, (3.2) holds when $n = 2k$. In a similar manner, we can prove that (3.2) holds when $n = 2k + 1$ too.

Let $0 < \epsilon < 1$. Let $\sum_{n=1}^{\infty} a_n = \sum_{n=1}^{\infty} \frac{(-1)^{n+1}}{n+1}$. Now, by using (3.2), we have

$$|a_{n+1} + a_{n+2} + \cdots + a_{n+p}|$$

$$= \left| \frac{1}{n+1} - \frac{1}{n+2} + \frac{1}{n+3} - \frac{1}{n+4} + \cdots + \frac{(-1)^{p+1}}{n+p} \right|$$

$$= \frac{1}{n+1} - \frac{1}{n+2} + \frac{1}{n+3} - \frac{1}{n+4} + \cdots + \frac{(-1)^{p+1}}{n+p}$$

$$< \frac{1}{n+1}.$$

If $n > \frac{1}{\epsilon} - 1$, i.e., $n + 1 > \frac{1}{\epsilon}$, i.e., $\frac{1}{n+1} < \epsilon$, then we have

$$|a_{n+1} + a_{n+2} + \cdots + a_{n+p}| < \frac{1}{n+1} < \epsilon.$$

Let N be a positive integer $> \frac{1}{\epsilon} - 1$. Then, for every $n \geq N$ and $p = 1, 2, \ldots$, we have

$$|a_{n+1} + a_{n+2} + \cdots + a_{n+p}| < \epsilon.$$

Now, appealing to Theorem 3.3.2, we conclude that $\sum_{n=1}^{\infty} \frac{(-1)^{n+1}}{n}$ converges (in fact, the series converges to $\log 2$. The reader should find it an interesting exercise).

Theorem 3.3.5. *Let $\sum_{n=1}^{\infty} a_n$ be a series of real numbers. Let*

$$p_n = \frac{|a_n| + a_n}{2}, \quad q_n = \frac{|a_n| - a_n}{2} \quad \text{for every } n = 1, 2, \ldots.$$

Then, we have the following:

(1) *If $\sum_{n=1}^{\infty} a_n$ is conditionally convergent, both $\sum_{n=1}^{\infty} p_n$ and $\sum_{n=1}^{\infty} q_n$ diverge;*

(2) *If $\sum_{n=1}^{\infty} a_n$ converges absolutely, both $\sum_{n=1}^{\infty} p_n$ and $\sum_{n=1}^{\infty} q_n$ converge and*

$$\sum_{n=1}^{\infty} a_n = \sum_{n=1}^{\infty} p_n - \sum_{n=1}^{\infty} q_n.$$

Proof. First, we note that

$$a_n = p_n - q_n, \quad |a_n| = p_n + q_n \quad \text{for every } n = 1, 2, \ldots.$$

(1) Let $\sum_{n=1}^{\infty} a_n$ be conditionally convergent, i.e., $\sum_{n=1}^{\infty} a_n$ converges and $\sum_{n=1}^{\infty} |a_n|$ diverges. If $\sum_{n=1}^{\infty} q_n$ converges, then $\sum_{n=1}^{\infty} p_n$ also converges since $p_n = a_n + q_n$ for every $n = 1, 2, \ldots$.

Similarly, if $\sum_{n=1}^{\infty} p_n$ converges, then $\sum_{n=1}^{\infty} q_n$ converges too since $q_n = p_n - a_n$ for every $n = 1, 2, \ldots$. Thus, if either $\sum_{n=1}^{\infty} p_n$ or $\sum_{n=1}^{\infty} q_n$ converges, then both converge and, consequently, $\sum_{n=1}^{\infty} |a_n|$ converges since $|a_n| = p_n + q_n$ for every $n = 1, 2, \ldots$, which is a contradiction and so we prove (1).

The proof of (2) is easy and left as an exercise to the reader. This completes the proof.

3.4 Convergence Tests for Infinite Series

Now, we introduce some convergence tests for infinite series.

Theorem 3.4.1. (The Comparison Test) *If $a_n, b_n > 0$ for every $n = 1, 2, \ldots$ and there exist $M > 0$ and a positive integer N such that*

$$a_n < Mb_n \quad \text{for ever } n \geq N, \tag{3.3}$$

then the convergence of $\sum_{n=1}^{\infty} b_n$ implies the convergence of $\sum_{n=1}^{\infty} a_n$.

Proof. Let a series $\sum_{n=1}^{\infty} b_n$ converge. Then, the sequence of partial sums of $\sum_{n=1}^{\infty} b_n$ is bounded and so the sequence of partial sums of $\sum_{n=1}^{\infty} a_n$ is bounded in view of (3.3). Since $a_n > 0$ for every $n = 1, 2, \ldots$, $\sum_{n=1}^{\infty} a_n$ converges by using Theorem 3.3.1. This completes the proof.

Theorem 3.4.2. (The Limit Comparison Test) *Let $a_n, b_n > 0$ for every $n = 1, 2, \ldots$ and*

$$\lim_{n \to \infty} \frac{a_n}{b_n} = 1. \tag{3.4}$$

Then, the following are equivalent:

(1) *A series $\sum_{n=1}^{\infty} a_n$ converges;*

(2) *A series $\sum_{n=1}^{\infty} b_n$ converges.*

Proof. In view of (3.4), corresponding to $\epsilon = \frac{1}{2}$, there exists a positive integer N such that

$$\left| \frac{a_n}{b_n} - 1 \right| < \frac{1}{2} \quad \text{for every } n \geq N,$$

which implies

$$-\frac{1}{2} < \frac{a_n}{b_n} - 1 < \frac{1}{2}, \quad \text{i.e.,} \quad \frac{1}{2} < \frac{a_n}{b_n} < \frac{3}{2}.$$

Now, applying Theorem 3.4.1 twice, we get the result. This completes the proof.

Remark 3.4.1. Theorem 3.4.2 holds if

$$\lim_{n \to \infty} \frac{a_n}{b_n} = c \ \text{ for any } c \neq 0.$$

It is worthwhile to see what happens when $c = 0$, which is left as an exercise to the reader.

Theorem 3.4.3. *The geometric series* $1 + x + x^2 + \cdots$ *converges to* $\frac{1}{1-x}$ *if* $|x| < 1$ *and diverges if* $|x| \geq 1$.

Proof. Observe that

$$(1 - x) \sum_{k=0}^{n} x^k = \sum_{k=0}^{n} (x^k - x^{k+1}) = 1 - x^{n+1}.$$

If $|x| < 1$, then $\displaystyle\lim_{n \to \infty} x^{n+1} = 0$, so that $\displaystyle\sum_{k=0}^{\infty} x^k$ converges to $\frac{1}{1-x}$ if $|x| < 1$.

However, if $|x| \geq 1$, then $x^n \nrightarrow 0$ as $n \to \infty$, and so the series $\displaystyle\sum_{k=0}^{\infty} x_k$ diverges in this case. This completes the proof.

Theorem 3.4.4. (The Ratio Test) *Let* $\sum_{n=1}^{\infty} a_n$ *be a series of non-zero real numbers. Let*

$$r = \liminf_{n \to \infty} \left| \frac{a_{n+1}}{a_n} \right|, \quad R = \liminf_{n \to \infty} \left| \frac{a_{n+1}}{a_n} \right|.$$

Then, we have the following:

(1) *The series* $\displaystyle\sum_{n=1}^{\infty} a_n$ *converges absolutely if* $R < 1$;

(2) *The series* $\displaystyle\sum_{n=1}^{\infty} a_n$ *diverges if* $r > 1$;

(3) *The test is in conclusive if* $r \leq 1 \leq R$.

Proof. (1) Let $R < 1$. Choose a real number x such that $R < x < 1$. Let $\epsilon = x - R$. Using the definition of $\limsup$, we can choose a positive integer N such that, for every $n \geq N$,

$$\left| \frac{a_{n+1}}{a_n} \right| < R + \epsilon = x = \frac{x^{n+1}}{x_n},$$

which implies

$$\frac{|a_{n+1}|}{x^{n+1}} < \frac{|a_n|}{x^n} \ \text{ for every } n \geq N.$$

Thus, we have

$$\frac{|a_n|}{x^n} \leq \frac{|a_N|}{x^N}, \quad \text{i.e., } \ |a_n| \leq M x^n \text{ for every } n \geq N,$$

where $M = \frac{|a_N|}{x^N}$. Using Theorems 3.4.1 and 3.4.2, it follows that $\sum_{n=1}^{\infty} a_n$ converges absolutely.

(2) If $r > 1$, then we can find a suitable positive integer N' such that

$$|a_{n+1}| > |a_n| \text{ for every } n \geq N'$$

and so we cannot have $\lim_{n \to \infty} a_n = 0$. Thus, $\sum_{n=1}^{\infty} a_n$ diverges in this case.

To prove (3), consider the series $\sum_{n=1}^{\infty} \frac{1}{n}$ and $\sum_{n=1}^{\infty} \frac{1}{n^2}$. In both cases, $r = R = 1$. However, $\sum_{n=1}^{\infty} \frac{1}{n}$ diverges and $\sum_{n=1}^{\infty} \frac{1}{n^2}$ converges (Prove it!). This completes the proof.

Theorem 3.4.5. (The Root Test) *Let $\sum_{n=1}^{\infty} a_n$ be a series of real numbers and*

$$\rho = \limsup_{n \to \infty} |a_n|^{\frac{1}{n}}.$$

Then, we have the following:

(1) $\sum_{n=1}^{\infty} a_n$ *converges absolutely if $\rho < 1$;*

(2) $\sum_{n=1}^{\infty} a_n$ *diverges if $\rho > 1$;*

(3) *The test is inconclusive if $\rho = 1$.*

Proof. (1) Let $\rho < 1$. Let x be a real number such that $\rho < x < 1$. Using the definition of lim sup, there exists a positive integer N such that

$$|a_n|^{\frac{1}{n}} < x, \text{ i.e., } |a_n| < x^n \text{ for every } n \geq N.$$

Using Theorems 3.4.3 and 3.4.1, it follows that $\sum_{n=1}^{\infty} a_n$ converges absolutely and so we prove (1).

To prove (2), let $\rho > 1$. Again, using the definition of lim sup, we have

$$|a_n|^{\frac{1}{n}} > 1, \text{ i.e., } |a_n| > 1 \text{ for infinity of } n's.$$

So, we cannot have $\lim_{n \to \infty} a_n = 0$. Consequently, it follows that $\sum_{n=1}^{\infty} a_n$ diverges in this case.

For (3), consider the series $\sum_{n=1}^{\infty} \frac{1}{n}$ and $\sum_{n=1}^{\infty} \frac{1}{n^2}$. For both of the series, $\rho = 1$, but the series $\sum_{n=1}^{\infty} \frac{1}{n^2}$ converges (why?) while the series $\sum_{n=1}^{\infty} \frac{1}{n}$ diverges (why?). This completes the proof.

In this context, it is worthwhile to mention that the Root Test is "more powerful" than the Ratio Test in the sense that there are cases where the Ratio Test fails while the Root Test succeeds.

3.5　Rearrangements of Series

Definition 3.5.1. Let f be a one-to-one function of $\mathbb{Z}^+$ onto $\mathbb{Z}^+$. Let $\sum_{n=1}^{\infty} a_n$ and $\sum_{n=1}^{\infty} b_n$ be two series such that

$$b_n = a_{f(n)} \text{ for every } n = 1, 2, \ldots. \tag{3.5}$$

Then, $\sum_{n=1}^{\infty} b_n$ is called a *rearrangement* of $\sum_{n=1}^{\infty} a_n$.

Note that (3.5) implies that

$$a_n = b_{f^{-1}(n)} \text{ for every } n = 1, 2, \ldots,$$

where f^{-1} is a one-to-one function of $\mathbb{Z}^+$ onto $\mathbb{Z}^+$, and so $\sum_{n=1}^{\infty} a_n$ is a rearrangement of $\sum_{n=1}^{\infty} b_n$, whenever $\sum_{n=1}^{\infty} b_n$ is a rearrangement of $\sum_{n=1}^{\infty} a_n$.

Theorem 3.5.1. *Let $\sum_{n=1}^{\infty} a_n$ be an absolutely convergent series with sum s. Then, every rearrangement of $\displaystyle\sum_{n=1}^{\infty} a_n$ also converges absolutely with sum s.*

Proof. Let $\{b_n\}$ be the sequence defined by (3.5). Now, we have

$$|b_1| + |b_2| + \cdots + |b_n| = |a_{f(1)}| + |a_{f(2)}| + \cdots + |a_{f(n)}|$$
$$\leq \sum_{k=1}^{\infty} |a_k|,$$

so that $\sum_{n=1}^{\infty} |b_n|$ has bounded partial sums. Consequently, $\sum_{n=1}^{\infty} |b_n|$ converges, i.e., $\sum_{n=1}^{\infty} b_n$ converges absolutely.

Now, we claim that $\sum_{n=1}^{\infty} b_n = s$. Let

$$s_n = \sum_{k=1}^{n} a_k, \quad t_n = \sum_{k=1}^{n} b_k \text{ for every } n = 1, 2, \ldots.$$

Since $\lim_{n \to \infty} s_n = s$, given $\epsilon > 0$, choose a positive integer N such that

$$|s_N - s| < \frac{\epsilon}{2}, \text{ i.e., } \left| s_N - \sum_{k=1}^{\infty} a_k \right| < \frac{\epsilon}{2}, \tag{3.6}$$

which implies

$$|a_{N+1} + a_{N+2} + \cdots| < \frac{\epsilon}{2}.$$

Now, using (3.6), we have

$$|t_n - s| = |(t_n - s_N) + (s_N - s)|$$
$$\leq |t_n - s_N| + |s_N - s|$$
$$< |t_n - s_N| + \frac{\epsilon}{2}.$$

Choose a positive integer M such that

$$\{1, 2, \ldots, N\} \subseteq \{f(1), f(2), \ldots, f(M)\},$$

so that $n > M$ implies that $f(n) > N$. For such an n, using (3.6), we have

$$
\begin{aligned}
|t_n - s_N| &= |(b_1 + b_2 + \cdots + b_n) - (a_1 + a_2 + \cdots + a_N)| \\
&= |(a_{f(1)} + a_{f(2)} + \cdots + a_{f(n)}) - (a_1 + a_2 + \cdots + a_N)| \\
&\leq |a_{N+1}| + |a_{N+2}| + \cdots \\
&< \frac{\epsilon}{2},
\end{aligned}
$$

where note that $a_1, a_2, \ldots, a_N$ cancel out in the subtraction. Thus, we have

$$|t_n - s_N| < \frac{\epsilon}{2} + \frac{\epsilon}{2} = \epsilon \text{ for every } n > M,$$

which implies

$$\lim_{n \to \infty} t_n = s, \text{ i.e., } \sum_{k=1}^{\infty} b_k = s.$$

This completes the proof.

3.6   Riemann's Theorem on Conditionally Convergent Series of Real Numbers

Riemann proved that any conditionally convergent series of real numbers can be suitably rearranged so that the rearranged series converges to a preassigned sum.

Theorem 3.6.1. *Let $\sum_{n=1}^{\infty} a_n$ be a conditionally convergent series of real numbers. Let $x \in (-\infty, +\infty)$. Then, there exists a rearrangement $\sum_{n=1}^{\infty} b_n$ of the series $\sum_{n=1}^{\infty} a_n$, which converges to x.*

Proof. We first observe that discarding the terms of a series which are 0 does not affect the behavior of the series. So, without loss of generality, we can suppose that $a_n \neq 0$; let p_n denote the nth positive term of $\sum_{n=1}^{\infty} a_n$ and q_n denote its nth negative term. Since $\sum_{n=1}^{\infty} a_n$ is conditionally convergent, $\sum_{n=1}^{\infty} p_n$ and $\sum_{n=1}^{\infty} q_n$ are divergent series of positive terms. Let $\{x_n\}$ be a sequence of real numbers such that $x_1 > 0$ and

$$\lim_{n \to \infty} x_n = x.$$

Now, take just enough number of positive terms (say) $k(1)$ such that

$$p_1 + p_2 + \cdots + p_{k(1)} > x$$

and just enough number of negative terms (say) $r(1)$ such that

$$p_1 + p_2 + \cdots + p_{k(1)} - q_1 - q_2 - \cdots - q_{r(1)} < x.$$

We next take just enough further positive terms (say) $k(w)$ such that

$$\begin{aligned} p_1 + p_2 + \cdots + p_{k(1)} - q_1 - q_2 - \cdots - q_{r(1)} \\ + p_{k(1)+1} + p_{k(1)+2} + \cdots + p_{k(2)} > x, \end{aligned}$$

which is followed by just enough further negative terms (say) $r(2)$ such that

$$\begin{aligned} p_1 + p_2 + \cdots + p_{k(1)} - q_1 - q_2 - \cdots - q_{r(1)} \\ + p_{k(1)+1} + p_{k(1)+2} + \cdots + p_{k(2)} \\ - q_{r(1)+1} - q_{r(1)+2} - \cdots - q_{r(2)} \\ < x. \end{aligned}$$

If we continue the above process, then such a choice of terms is possible since $\sum_{n=1}^{\infty} p_n$ and $\sum_{n=1}^{\infty} q_n$ are divergent series of positive terms. Thus, we obtain a rearrangement $\sum_{n=1}^{\infty} b_n$ and $\sum_{n=1}^{\infty} a_n$. We can prove that the sequence of partial sums of this rearrangement $\sum_{n=1}^{\infty} b_n$ converges to x, i.e., $\sum_{n=1}^{\infty} b_n$ converges to x. This completes the proof.

Remark 3.6.1. Theorem 3.6.1 holds if $x = +\infty$ or $-\infty$.

3.7 Cauchy Multiplications of Series

Definition 2.7.1. Given two series $\sum_{n=1}^{\infty} a_n$ and $\sum_{n=1}^{\infty} b_n$ of real numbers, let

$$c_n = \sum_{k=0}^{n} a_k b_{n-k} \quad \text{for every } n = 0, 1, 2, \ldots. \tag{3.7}$$

The series $\sum_{n=0}^{\infty} c_n$ is called the *Cauchy product* of $\sum_{n=0}^{\infty} a_n$ and $\sum_{n=0}^{\infty} b_n$.

Theorem 3.7.1. (Mertens' Theorem) *Suppose that a series $\sum_{n=0}^{\infty} a_n$ converges absolutely and has the sum A and $\sum_{n=0}^{\infty} b_n$ converges and has the sum B. Then, the Cauchy product of the two series converges and has the sum AB.*

Proof. Let

$$A_n = \sum_{k=0}^{n} a_k, \quad B_n = \sum_{k=0}^{n} b_k, \quad C_n = \sum_{k=0}^{n} c_k \quad \text{for every } n = 0, 1, 2, \ldots,$$

where c_n is defined by (3.7). Let

$$d_n = B - B_n, \quad e_n = \sum_{k=0}^{n} a_k d_{n-k} \quad \text{for every } n = 0, 1, 2, \ldots.$$

Now, we have

$$
\begin{aligned}
C_p = \sum_{n=0}^{p} c_n &= \sum_{n=0}^{p} \sum_{k=0}^{n} a_k b_{n-k} \\
&= \sum_{k=0}^{p} \sum_{n=k}^{p} a_k b_{n-k} \\
&= \sum_{k=0}^{p} a_k \sum_{n=k}^{p} b_{n-k} \\
&= \sum_{k=0}^{p} a_k (b_0 + b_1 + \cdots + b_{p-k}) \\
&= \sum_{k=0}^{p} a_k B_{p-k} \\
&= \sum_{k=0}^{p} a_k (B - d_{p-k}) \\
&= B A_p - e_p.
\end{aligned}
\tag{3.8}
$$

It now suffices to prove that $e_p \to 0$ as $p \to \infty$. Since $\lim_{n \to \infty} B_n = B$, it follows that $\{d_n\}$ converges to 0 and so $\{d_n\}$ is bounded. So, there exists $M > 0$ such that

$$|d_n| \le M \quad \text{for every } n = 0, 1, 2, \ldots. \tag{3.9}$$

Let $\sum_{n=0}^{\infty} |a_n| = K$. For every $\epsilon > 0$, we can choose a positive integer N such that, for every $n > N$,

$$|d_n| < \frac{\epsilon}{2K} \tag{3.10}$$

and

$$\sum_{n=N+1}^{\infty} |a_n| < \frac{\epsilon}{2M}. \tag{3.11}$$

Now, it follows from (3.9), (3.10), and (3.11) that, for every $p > 2N$,

$$
\begin{aligned}
|e_p| &= \left| \sum_{k=0}^{p} a_k d_{p-k} \right| \\
&= \left| \sum_{k=0}^{N} a_k d_{p-k} + \sum_{k=N+1}^{p} a_k d_{p-k} \right| \\
&\leq \sum_{k=0}^{N} |a_k d_{p-k}| + \sum_{k=N+1}^{p} |a_k d_{p-k}| \\
&< \frac{\epsilon}{2K} \sum_{k=0}^{N} |a_k| + M \sum_{k=N+1}^{p} |a_k| \\
&< \frac{\epsilon}{2K} \sum_{k=0}^{\infty} |a_k| + M \sum_{k=N+1}^{\infty} |a_k| \\
&< \frac{\epsilon}{2K} K + M \frac{\epsilon}{2M} \\
&= \epsilon.
\end{aligned}
$$

Taking limit as $p \to \infty$ in (3.8), we have

$$
\lim_{p \to \infty} C_p = AB,
$$

i.e., $\sum_{n=0}^{\infty} c_n$ converges, and its sum is AB. This completes the proof.

3.8 Exercises

Excercise 3.1 Let $\{a_n\}$ be a sequence of real numbers. Prove that

(1) $\liminf_{n \to \infty} a_n \leq \limsup_{n \to \infty} a_n$;

(2) The sequence $\{a_n\}$ converges if and only if $\liminf_{n \to \infty} a_n$ and $\limsup_{n \to \infty} a_n$ exist finitely and are equal, in which case, $\liminf_{n \to \infty} a_n = \limsup_{n \to \infty} a_n = \lim_{n \to \infty} a_n$;

(3) The sequence $\{a_n\}$ diverges to $+\infty$ if and only if $\liminf_{n \to \infty} a_n = \limsup_{n \to \infty} a_n = +\infty$;

(4) The sequence $\{a_n\}$ diverges to $-\infty$ if and only if $\liminf_{n \to \infty} a_n = \limsup_{n \to \infty} a_n = -\infty$.

Excercise 3.2 Let $a_n \leq b_n$ for every $n = 1, 2, \ldots$. Prove that

(1) $\liminf\limits_{n\to\infty} a_n \leq \liminf\limits_{n\to\infty} b_n$;

(2) $\limsup\limits_{n\to\infty} a_n \leq \limsup\limits_{n\to\infty} b_n$.

Excercise 3.3 Find $\liminf\limits_{n\to\infty} a_n$ and $\limsup\limits_{n\to\infty} a_n$, when

(1) $a_n = (-1)^n$ for every $n = 1, 2, \ldots$;

(2) $a_n = (-1)^n n$ for every $n = 1, 2, \ldots$;

(3) $a_n = (-1)^n \left(1 + \frac{1}{n}\right)$ for every $n = 1, 2, \ldots$.

Excercise 3.4 Let $\sum\limits_{n=1}^{\infty} a_n = A$ and $\sum\limits_{n=1}^{\infty} b_n = B$. Prove that $\sum\limits_{n=1}^{\infty} (\alpha a_n + \beta b_n) = \alpha A + \beta B$, where α and β are constants.

Excercise 3.5 (1) Let $a_n > a$ for every $n = 1, 2, \ldots$. Prove that

$$\liminf\limits_{n\to\infty} \frac{a_{n+1}}{a_n} \leq \liminf\limits_{n\to\infty} a_n^{\frac{1}{n}} \leq \limsup\limits_{n\to\infty} a_n^{\frac{1}{n}} \leq \limsup\limits_{n\to\infty} \frac{a_{n+1}}{a_n};$$

(2) Let $a_n = \frac{n^n}{n!}$ for every $n = 1, 2, \ldots$. Prove that $\lim\limits_{n\to\infty} \frac{a_{n+1}}{a_n} = e$ and deduce that

$$\lim\limits_{n\to\infty} \frac{n}{(n!)^{\frac{1}{n}}} = e.$$

Excercise 3.6 (1) Prove that a monotonic increasing sequence which is bounded above converges to its sup;

(2) Prove also that a monotonic decreasing sequence which is bounded below converges to its inf.

Excercise 3.7 Prove that the sequence $(1 + \frac{1}{n})^n$ converges. (Hint: Use Exercise 3.6.)

Excercise 3.8 Prove that the sequence $\{x_n\}$ defined by

$$x_1 = \sqrt{2}, \quad x_{n+1} = \sqrt{2x_n} \text{ for every } n = 1, 2, \ldots$$

converges to 2. (Hint: Use Exercise 3.6.)

Excercise 3.9 (1) Prove that the sequence $\{x_n\}$ defined by

$$x_1 = \sqrt{7}, \quad x_{n+1} = \sqrt{7 + x_n} \text{ for every } n = 1, 2, \ldots$$

converges.

(2) Find their limits too.

Excercise 3.10 Using Cauchy's General Principle of the convergence, prove that the sequence $\{1 + \frac{1}{2} + \cdots + \frac{1}{n}\}$ does not converge.

Excercise 3.11 Prove that the following sequences $\{x_n\}$ are convergent, where

(1) $x_n = \frac{1}{1!} + \frac{1}{2!} + \cdots + \frac{1}{n!}$;

(2) $x_n = \frac{1}{n+1} + \frac{1}{n+2} + \cdots + \frac{1}{2n}$.

Excercise 3.12 Find $\lim\limits_{n \to \infty} x_n$, where

(1) $x_n = \frac{1}{\sqrt{n^2+1}} + \frac{1}{\sqrt{n^2+2}} + \cdots + \frac{1}{\sqrt{n^2+n}}$; (Ans. 1)

(2) $x_n = \frac{1}{n^2} + \frac{1}{(n+1)^2} + \cdots + \frac{1}{(2n)^2}$. (Ans. 0)

Excercise 3.13 [**Cesàro's Theorem**] If $\lim\limits_{n \to \infty} a_n = \ell$, then prove that

$$\lim_{n \to \infty} \frac{a_1 + a_2 + \cdots + a_n}{n} = \ell.$$

Excercise 3.14 Test for convergence of the following series, p, q being fixed real numbers:

(1) $\sum\limits_{n=1}^{\infty} n^3 e^{-n}$;

(2) $\sum\limits_{n=2}^{\infty} (\log n)^p$;

(3) $\sum\limits_{n=1}^{\infty} n^{-1-\frac{1}{n}}$;

(4) $\sum\limits_{n=1}^{\infty} (\sqrt{1+n^2} - n)$;

(5) $\sum\limits_{n=1}^{\infty} (n^{\frac{1}{n}} - 1)^n$;

(6) $\sum\limits_{n=2}^{\infty} \frac{1}{n^p - n^q}$, where $0 < q < p$;

(7) $\sum\limits_{n=1}^{\infty} \frac{1}{n \log(1 + \frac{1}{n})}$;

(8) $\sum\limits_{n=2}^{\infty} \frac{1}{(\log n)^{\log n}}$.

Excercise 3.15 Let $\sum_{n=1}^{\infty} a_n$ be an absolutely convergent series of real numbers. Prove that

(1) $\sum_{n=1}^{\infty} a_n^2$ converges absolutely;

(2) $\sum_{n=1}^{\infty} \dfrac{a_n}{1+a_n}$, where $a_n \neq -1$ for every $n = 1, 2, \ldots$ also converges absolutely.

Excercise 3.16 Let $\{a_n\}$ and $\{b_n\}$ be two sequences of real numbers. Let $A_n = \sum_{k=1}^{n} a_k$ for every $n = 1, 2, \ldots$. Prove that

$$\sum_{k=1}^{n} a_k b_k = A_n b_{n+1} - \sum_{k=1}^{n} A_k(b_{k+1} - b_k) \quad \text{for every } n = 1, 2, \ldots.$$

Excercise 3.17 [Dirichlet's Test] Let $\sum_{n=1}^{\infty} a_n$ be a series of real numbers such that $\{A_n\}$ is bounded, where $A_n = \sum_{k=1}^{n} a_k$ for every $n = 1, 2, \ldots$. Let $\{b_n\}$ be a decreasing sequence converging to 0. Prove that $\sum_{n=1}^{\infty} a_n b_n$ converges. (Hint: Use Exercise 8.)

Excercise 3.18 [Abel's Test] Prove that, if $\sum_{n=1}^{\infty} a_n$ converges and $\{b_n\}$ is a monotonic convergent sequence, then $\sum_{n=1}^{\infty} a_n b_n$ converges.

Excercise 3.19 Prove that, if $s \sum_{n=1}^{\infty} a_n$ converges and $\sum_{n=1}^{\infty} (b_n - b_{n+1})$ converges absolutely, then $\sum_{n=1}^{\infty} a_n b_n$ converges.

Excercise 3.20 Prove that, if $\sum_{n=0}^{\infty} a_n$ and $\sum_{n=0}^{\infty} b_n$ are absolutely convergent series of real numbers, then $\sum_{n=0}^{\infty} c_n$ is also absolutely convergent and

$$\sum_{n=0}^{\infty} c_n = \left(\sum_{n=0}^{\infty} a_n \right) \left(\sum_{n=0}^{\infty} b_n \right),$$

where $c_n = \sum_{k=0}^{n} a_k b_{n-k}$ for every $n = 0, 1, 2, \ldots$.

4

Metric Spaces – Basic Concepts, Complete Metric Spaces

In this chapter, we give the definition of metric spaces and its basic concepts, point set topology in metric spaces, some convergences theorems of sequences in metric spaces. Finally, we give the definition of Cauchy sequences in metric spaces and complete metric spaces.

4.1 Metric and Metric Spaces

Definition 4.1.1. By a *metric* on a non-empty set M, we mean a mapping $d : M \times M \to \mathbb{R}$ such that, for every $x, y, z \in M$,

(M1) $d(x, y) \geq 0$ and $d(x, y) = 0$ if and only if $x = y$;

(M2) $d(x, y) = d(y, x)$;

(M3) $d(x, y) \leq d(x, z) + d(z, y)$ (*triangle inequality*).

The pair (M, d) is called a *metric space*.

Here, $d(x, y)$ can be thought of as the distance from x to y for every $x, y \in M$. Note that we can define more than one metric on the non-empty set M. If d_1, d_2 are metrices defined on M, then the metric space (M, d_1) is usually different from the metric space (M, d_2), though the underlying set M is the same.

Example 4.1.1. (1) Let $M = \mathbb{R}$. For every $x, y \in \mathbb{R}$, define a mapping $d : M \times M \to \mathbb{R}$ by
$$d(x, y) = |x - y|.$$

Then, d is a metric. It is called the *Euclidean metric* on $\mathbb{R}$. Whenever we refer to $\mathbb{R}$, it is tacitly assumed that the metric is the Euclidean metric unless otherwise stated.

  (2) Let $M = \mathbb{C}$, the complex plane. For every $z_1, z_2 \in \mathbb{C}$, define a mapping $d : M \times M \to \mathbb{C}$ by
$$d(z_1, z_2) = |z_1 - z_2|.$$

d is a metric on $\mathbb{C}$. In this context, note that $\mathbb{C}$ and the Euclidean space $\mathbb{R}^2$ are indistinguishable in the sense that $\mathbb{C}$ and $\mathbb{R}^2$ have the same points and the same metric.

(3) Let M be any non-empty set. For every $x, y \in M$, define a mapping $d : M \times M \to \mathbb{R}$ by

$$d(x, y) = \begin{cases} 1, & \text{if } x \neq y; \\ 0, & \text{if } x = y. \end{cases}$$

We can check that d is a metric on M. It is called the *discrete metric* and (M, d) is called the *discrete metric space*;

(4) Refer to Example (2). We take $M = \mathbb{R}^2$. For every $x = (x_1, x_2)$, $y = (y_1, y_2) \in \mathbb{R}^2$, define a mapping $d : M \times M \to \mathbb{R}^2$ by

$$d(x, y) = \sqrt{(x_1, y_1)^2 + 4(x_2 - y_2)^2}.$$

Then, d defines a metric on M. Note that the metric space $\mathbb{R}^2$ with the Euclidean metric is different from the metric space $\mathbb{R}^2$ with the above metric.

The reader should try to construct more examples of metric spaces.

4.2   Point Set Topology in Metric Spaces

Definition 4.2.1. Let (M, d) be a metric space. The *open ball* $B(a; r)$ with centre a and radius $r > 0$ is defined by

$$B(a; r) = \{x \in M : d(x, a) < r\}.$$

Sometimes, $B(a; r)$ is denoted by $B_M(a; r)$ to emphasize that its points are in M. Let $S \subseteq M$. If (S, d) itself is a metric space, S is called a *metric subspace* of M. In such a case,

$$B_S(a; r) = B_M(a; r) \bigcap S.$$

Example 4.2.1. Let $M = \mathbb{R}$. Then, we have

$$B_M(0, 1) = (-1, 1).$$

In the metric subspace $S = [0, 1]$, we have

$$B_S(0, 1) = B_M(0, 1) \bigcap S$$
$$= (-1, 1) \bigcap [0, 1]$$
$$= [0, 1).$$

Definition 4.2.2. Let (M, d) be a metric space and $S \subseteq M$.

(1) A point $a \in S$ is called an *interior point* of S if there exists an open ball $B_M(a; r)$ such that

$$B_M(a; r) \subseteq S.$$

The set of all interior points of S is called the *interior* of S, denoted by *int S*;

(2) S is said to be *open* in M if

$$int\, S = S,$$

i.e., if all the points of S are interior points of S;

(3) S is said to be *closed* in M if $M - S$ is open in M.

Note that every open ball $B_M(a; r)$ in a metric space M is an open set in M. In a discrete metric space M, every subset S of M is open, in fact, if $x \in S$, then we have

$$B\left(x; \frac{1}{2}\right) = \{x\} \subseteq S.$$

Consequently, every subset S of M is closed too. Again, in the metric subspace $S = [0, 1]$ of $\mathbb{R}$, every interval of the form $[0, a), (a, 1], 0 < a < 1$ is an open set in S, while they are not open in $\mathbb{R}$. This shows that if S is a metric subspace of M, sets which are open in S need not be open in M.

The following result establishes the relationship of open sets in S and open sets in M:

Theorem 4.2.1. *Let (S, d) be a metric subspace of (M, d) and $X \subseteq S$. Then, the following are equivalent:*

(1) *X is open in S;*

(2) *$X = Y \cap S$ for some set Y which is open in M.*

Proof. (1) $\implies$ (2) Let Y be open in M and $X = Y \cap S$. We claim that X is open in S. Let $x \in X$ and so $x \in Y$. Since Y is open in M, we have $B_M(x; r) \subseteq Y$ for some $r > 0$. Now, it follows that

$$B_S(x; r) = B_M(x; r) \bigcap S \subseteq Y \bigcap S = X,$$

so that X is open in S.

(2) $\implies$ (1) Let X be open in S. Now, we prove that

$$X = Y \bigcap S$$

for some open set Y in M. Now, for every $x \in X$, there is an open ball

$$B_S(x; r_x) \subseteq X.$$

Note that

$$B_S(x; r_x) = B_M(x; r_x) \bigcap S.$$

Let

$$Y = \bigcup_{x \in X} B_M(x; r_x).$$

Then, Y is open in M and

$$Y \bigcap S = X.$$

This completes the proof.

Now, we have a similar result for closed sets.

Theorem 4.2.2. *Let (S, d) be a metric subspace of (M, d) and $X \subseteq S$. Then, the following are equivalent:*

(1) X *is closed in* S;

(2) $X = Y \bigcap S$ *for some set* Y *which is closed in* M.

Proof. $(2) \Longrightarrow (1)$ Let

$$X = Y \bigcap S,$$

where Y is closed in M. Then, $Z = M - Y$ is open in M. So, we have

$$Y = M - Z,$$

where Z is open in M. Thus, we have

$$X = S \bigcap Y$$
$$= S \bigcap (M - Z)$$
$$= S - (S \bigcap Z).$$

Since Z is open in M, $S \cap Z$ is open in S by using Theorem 4.2.1, and so $S - (S \cap Z)$ is closed in S, i.e., X is closed in S.

$(1) \Longrightarrow (2)$ Let X be closed in S. Let $Y = S - X$, and so Y is open in S. In view of Theorem 4.2.1, we have

$$Y = Z \bigcap S,$$

where Z is open in M. Now, it follows that

$$X = S - Y = S - (Z \bigcap S)$$
$$= S - Z$$
$$= S \bigcap (M - Z)$$
$$= S \bigcap A,$$

where $A = M - Z$ is close in M since Z is open in M. This completes the proof.

Definition 4.2.3. Let (M, d) be a metric space. Let $S \subseteq M$ and $x \in M$.

(1) The point x is called an *adherent point* of S if every open ball $B_M(x; r)$ with x as center contains at least one point of S;

(2) If every open ball $B_M(x; r)$ contains at least one point of S other than x, then x is called an *accumulation point* of S;

(3) The set of all adherent points of S is called the *closure* of S, denoted by $\overline{S}$;

(4) The set of all accumulation points of S is called the *derived set* of S, denoted by S';

(5) If $x \in S$ and x is not an accumulation point called an *isolated point* of S.

Remark 4.2.1. Note that
$$S \bigcup S' = \overline{S}.$$

Theorem 4.2.3. *Let (M, d) be a metric space M. Then, we have the following:*

(1) *The union of any collection of open sets is open;*

(2) *The intersection of any finite collection of open sets is open.*

Proof. (1) Let $\mathscr{F}$ be a collection of open sets and $S = \bigcup_{A \in \mathscr{F}} A$. Let $x \in S$. Then, $x \in A$ for some $A \in \mathscr{F}$. Since A is open, there exists an open ball $B_M(x; r) \subseteq A$. Since $A \subseteq S$, we have $B_M(x; r) \subseteq S$, i.e., x is an interior point of S for every $x \in S$, i.e., S is open.

(2) Let $S = \bigcap_{k=1}^{n} A_k$, where A_k is open for every $k = 1, 2, \ldots, n$. If $S = \emptyset$, we are through. Let $S \neq \emptyset$ and $x \in S$. So, $x \in A_k$ for every $k = 1, 2, \ldots, n$. Consequently, there exists an open ball

$$B_M(x; r_k) \subseteq A_k \text{ for every } k = 1, 2, \ldots, n.$$

Let $r = \min\{r_1, r_2, \ldots, r_n\}$. Then, we have

$$x \in B_M(x; r) \subseteq S,$$

i.e., x is an interior point of S for every $x \in S$, i.e., S is open. This completes the proof.

Remark 4.2.2. Theorem 4.2.3 shows that from given open sets, new open sets can be formed by taking arbitrary unions and finite intersections. However, arbitrary intersections may not lead to open sets as the following example illustrates:

$\left(-\frac{1}{n}, \frac{1}{n}\right)$ is open for every $n = 1, 2, \ldots$. But $\bigcap_{n=1}^{\infty} \left(-\frac{1}{n}, \frac{1}{n}\right) = \{0\}$, which is not an open set (why?).

Using Theorem 4.2.3, we have the following results for closed Sets.

Theorem 4.2.4. *In a metric space M, the intersection of any collection of closed sets is closed, while the union of any finite collection of closed sets is closed. Arbitrary union of closed sets need not be closed.*

The following result shows that a set cannot have an accumulation point unless it contains infinitely many points:

Theorem 4.2.5. *Let (M, d) be a metric space and $S \subseteq M$. If x is an accumulation point of S, then every open ball $B_M(x; r)$ contains infinitely many points of S different from x.*

Proof. Suppose that the conclusion is false, i.e., there exists an open ball $B_M(x; r)$, which contains only a finite number of points of S different from x, (say) $a_1, a_2, \ldots, a_n$. Let

$$r = \min_{1 \leq i \leq n} d(x, a_i).$$

Then, $B_m\left(x; \frac{r}{2}\right)$ is an open ball with x as center, which contains no points of S different from x. This contradicts the fact that x is an accumulation point of S. This completes the proof.

Now, we describe closed sets in several ways.

Theorem 4.2.6. *The following are equivalent:*

(1) *A set S in a metric space (M, d) is closed;*

(2) *It contains all its adherent points.*

Proof. $(1) \implies (2)$ Let S be closed and x be an adherent point of S. Now, we claim that $x \in S$. Suppose $x \notin S$. Then, $x \in M - S$. Since S is close, $M - S$ is open and so

$$B_M(x; r) \subseteq M - S \quad \text{for some} \quad r > 0.$$

Thus, $B_M(x; r)$ does not contain any point of S, which contradicts the fact that x is an adherent point of S. Consequently, $x \in S$, i.e., S contains all its adherent points.

$(2) \implies (1)$ Let S contain all its adherent points. Now, we prove that S is closed. Let $x \in M - S$. Then, $x \notin S$ and so x cannot be an adherent point of S by our assumption. So, for some $r > 0$, we have

$$B_M(x; r) \bigcap S = \emptyset.$$

Thus, we have

$$B_M(x; r) \subseteq M - S, \quad \text{i.e., } M - S \text{ is open,}$$

which implies that S is closed. This completes the proof.

Theorem 4.2.7. *A set S in a metric space (M, d) is closed if and only if $S = \overline{S}$.*

Proof. We first note that $S \subseteq \overline{S}$ always. In view of Theorem 4.2.6, S is closed if and only if $\overline{S} \subseteq S$. Consequently, S is closed if and only if $S = \overline{S}$. This completes the proof.

Theorem 4.2.8. *A set S in a metric space (M, d) is closed if and only if it contains all its accumulation points.*

Proof. We always have

$$\overline{S} = S \bigcup S'.$$

Using Theorem 4.2.7, S is closed if and only if $S' \subseteq S$. This completes the proof.

We can now reformulate Theorems 4.2.6–4.2.8 as follows:

Theorem 4.2.9. *For any set S in a metric space (M, d), the following are equivalent:*

(1) *S is closed;*

(2) *S contains all its adherent points;*

(3) *S contains all its accumulation points;*

(4) *$\overline{S} = S$.*

We recall that a set $S \subseteq \mathbb{R}$ is *bounded* if

$$S \subseteq [-a, a] \text{ for some } a > 0.$$

Now, we prove an important theorem.

Theorem 4.2.10. (The Bolzano–Weierstrass Theorem) *If a bounded set S in $\mathbb{R}$ contains infinitely many points, then there is at least one point in $\mathbb{R}$, which is an accumulation point of S.*

Proof. Since S is bounded, we have

$$S \subseteq [-a, a] \text{ for some } a > 0.$$

Since S contains infinitely many points, at least one of the subintervals $[-a, 0]$, $[0, a]$ contains infinity of points of S. Call one such subinterval $[a_1, b_1]$. Again, bisect $[a_1, b_1]$, and we get a subinterval $[a_2, b_2]$ containing infinity of points of S. Continuing this process, we get a countable collection of subintervals, the nth interval being $[a_n, b_n]$ of length

$$b_n - a_n = \frac{a}{2^{n-1}} \text{ for every } n = 1, 2, \dots.$$

Note that

$$\sup_{n \geq 1} a_n = \inf_{n \geq 1} b_n = x \, (\text{say}).$$

Now, we claim that x is an accumulation point of S. Let $r > 0$. Since

$$\lim_{n \to \infty} (b_n - a_n) = 0,$$

choose a positive integer N such that

$$b_n - a_n < \frac{r}{2} \text{ for every } n \geq N.$$

So, we have

$$b_N - a_N < \frac{r}{2}$$

and so

$$(a_N, b_N) \subseteq B(x; r).$$

Since (a_N, b_N) contains infinity of points of S, $B(x; r)$ contains a point of S other than x. Hence, x is an accumulation point of S. This completes the proof.

As an application of Theorem 4.2.10, we obtain the following result:

Theorem 4.2.11. (Cantor's Intersection Theorem) *Let $Q_1, Q_2, \ldots$ be a countable collection of non-empty sets in $\mathbb{R}$ such that*

(a) $Q_{k+1} \subseteq Q_k$ *for every* $k = 1, 2, \ldots$;

(b) *each Q_k is closed and Q_1 is bounded.*

Then, $\bigcap_{k=1}^{\infty} Q_k$ *is closed and non-empty.*

Proof. Let $S = \bigcap_{k=1}^{\infty} Q_k$. In view of Theorem 4.2.4, S is closed. If Q_{n_0} contains a finite number of points for some n_0, then the proof is trivial. So, let us suppose that every Q_k contains infinitely many points. Let us now form a collection of distinct points,

$$A = \{x_1, x_2, \ldots\}, \quad x_k \in Q_k \text{ for every } k = 1, 2, \ldots.$$

Then, $A \subseteq Q_1$, where Q_1 is bounded, and so A is bounded. Thus, A is an infinite and bounded subset of $\mathbb{R}$. Using Theorem 4.2.10, A has an accumulation point, (say) x.

Now, we claim that $x \in Q_k$ for every $k = 1, 2, \ldots$. Since Q_k is closed for every $k = 1, 2, \ldots$, it suffices to prove that x is an accumulation point of each Q_k. Since $Q_{k+1} \subseteq Q_k$, we have

$$x_i \in Q_k \text{ for every } i \geq k.$$

Since x is an accumulation point of A, every open ball with x as center contains infinitely many points of A, and so it contains infinitely many points of Q_k as well. Hence, x is an accumulation point of each Q_k. Consequently, we have $x \in Q_k$ for every $k = 1, 2, \ldots$, i.e., $x \in \bigcap_{k=1}^{\infty} Q_k$, i.e., $\bigcap_{k=1}^{\infty} Q_k \neq \emptyset$. This completes the proof.

4.3 Convergent and Divergent Sequences in a Metric Space

Definition 4.3.1. (1) A sequence $\{x_n\}$ of points in a metric space (M, d) is said to *converge* if there exists a point p in M with the following property: for every $\epsilon > 0$, there exists a positive integer N such that

$$d(x_n, p) < \epsilon \text{ whenever } n \geq N;$$

(2) We also say that $\{x_n\}$ converges to p and write $x_n \to p$ as $n \to \infty$;

(3) If no such p in M exists, then we say that the sequence $\{x_n\}$ *diverges*.

Theorem 4.3.1. *A sequence $\{x_n\}$ in a metric space (M, d) converges to at most one point in M.*

Proof. Let $\lim\limits_{n \to \infty} x_n = p$ and $\lim\limits_{n \to \infty} x_n = q$. We claim that $p = q$. Using the triangle inequality, we have

$$0 \leq d(p, q) \leq d(p, x_n) + d(x_n, q).$$

Taking the limit as $n \to \infty$ in the above inequality, we get

$$0 \leq d(p, q) \leq 0 + 0 = 0$$

since $d(p, x_n)$, $d(x_n, q) \to 0$ as $n \to \infty$. Thus, $d(p, q) = 0$, from which we get $p = q$. This completes the proof.

Theorem 4.3.1 shows that if a sequence $\{x_n\}$ converges, then the unique point to which it converges is called the limit of the sequence $\{x_n\}$, written as $\lim\limits_{n \to \infty} x_n$.

Theorem 4.3.2. *Let $\{x_n\}$ be a convergent sequence in a metric space (M, d). Then, $\{x_n\}$ is bounded in the sense that there exists $M > 0$ such that*

$$x_n \in B_M(p; r) \quad \text{for every } n = 1, 2, \ldots,$$

where $\lim\limits_{n \to \infty} x_n = p$.

Proof. Since $\lim\limits_{n \to \infty} x_n = p$, corresponding to $\epsilon = 1$, there exists a positive integer N such that

$$d(x_n, p) < 1 \quad \text{for every } n \geq N.$$

Let $r = 1 + \max\{d(x_1, p), d(x_2, p), \ldots, d(x_{N-1}, p)\}$. Then, we have

$$d(x_n, p) < r, \text{ i.e., } x_n \in B_M(p; r) \quad \text{for every } n = 1, 2, \ldots,$$

which implies that $\{x_n\}$ is bounded. This completes the proof.

Note that if T is the range of the sequence $\{x_n\}$, then p is an adherent point of T. Further, if T is infinite, p is an accumulation point of T.

The converse of the above statement is true and is very useful in analysis.

Theorem 4.3.3. *Let (M, d) be a metric space and $S \subseteq M$. If $p \in M$ is an adherent point of S, then there exists a sequence $\{x_n\}$ of points in S, which converges to p.*

Proof. Since p is an adherent point of S, the open ball $B_M\left(p, \frac{1}{n}\right)$ contains a point of S, (say) x_n, this being true for every $n = 1, 2, \ldots$, i.e.,

$$d(x_n, p) < \frac{1}{n} \quad \text{for every } n = 1, 2, \ldots.$$

Thus, $d(x_n, p) \to 0$ as $n \to \infty$, i.e., $\lim\limits_{n \to \infty} x_n = p$. This completes the proof.

The following is an important result:

Theorem 4.3.4. *Let (M, d) be a metric space. Then, the following are equivalent:*

(1) *A sequence $\{x_n\}$ converges to p;*

(2) *Every subsequence $\{x_{k(n)}\}$ of $\{x_n\}$ converges to p.*

Proof. $(2) \implies (1)$ Since $\{x_n\}$ is a subsequence of itself, the if part of the theorem is trivial.

$(1) \implies (2)$ Let $\lim\limits_{n \to \infty} x_n = p$. So, for every $\epsilon > 0$, there exists a positive integer N such that

$$d(x_n, p) < \epsilon \quad \text{for every } n \geq N.$$

Let $\{x_{k(n)}\}$ be a subsequence of $\{x_n\}$. Then, there exists a positive integer M such that $k(n) \geq N$ for every $n \geq M$, so that

$$d(x_{k(n)}, p) < \epsilon \quad \text{for every } n \geq M,$$

which implies $x_{k(n)} \to p$ as $n \to \infty$. This completes proof.

4.4 Cauchy Sequences and Complete Metric Spaces

If a sequence $\{x_n\}$ in a metric space (M, d) converges to p, then the terms of the sequence are ultimately close to p and so close to each other. This fact is brought out by the following result.

Theorem 4.4.1. *Let the sequence $\{x_n\}$ in a metric space (M, d) converge. Then, for every $\epsilon > 0$, there exists a positive integer N such that*

$$d(x_m, x_n) < \epsilon \quad \text{whenever } m, n \geq N.$$

Proof. Let $\lim_{n \to \infty} x_n = p$. So, given $\epsilon > 0$, there exists a positive integer N such that

$$d(x_n, p) < \frac{\epsilon}{2} \quad \text{whenever } n \geq N.$$

So, it follows that whenever $m, n \geq N$,

$$d(x_m, x_n) \leq d(x_m, p) + d(p, x_n)$$
$$< \frac{\epsilon}{2} + \frac{\epsilon}{2}$$
$$= \epsilon.$$

This completes the proof.

Theorem 4.4.1 is the motivation for the next definition:

Definition 4.4.1. A sequence $\{x_n\}$ in a metric space (M, d) is called a *Cauchy sequence* if, given any $\epsilon > 0$, there exists a positive integer N such that

$$d(x_m, x_n) < \epsilon \quad \text{whenever } m, n \geq N.$$

Theorem 4.4.1 states that every convergent sequence in a metric space is a Cauchy sequence. However, the converse is not true in general. For instance, consider the Euclidean subspace $S = (0, 1]$ of $\mathbb{R}$. The sequence $\{\frac{1}{n}\}$ is a Cauchy sequence in S, and this sequence does not converge to a point in S. The converse is true in the Euclidean space $\mathbb{R}$.

Theorem 4.4.2. *In the Euclidean space $\mathbb{R}$, every Cauchy sequence converges.*

Proof. Let $\{x_n\}$ be a Cauchy sequence in $\mathbb{R}$. Let T be the range of the sequence $\{x_n\}$. Now, we consider two cases as follows:

Case (1) If T is finite, then all except a finite number of terms of $\{x_n\}$ are equal (say) to p, and so $\{x_n\}$ converges to p.

Case (2) Suppose that T is infinite. Now, we claim that T has an accumulation point p. We first prove that T is bounded. Since $\{x_n\}$ is a Cauchy sequence, corresponding to $\epsilon = 1$, there exists a positive integer N such that

$$|x_n - x_N| < 1, \text{ i.e., } x_n \in B(x_N; 1) \text{ for every } n \geq N.$$

Let $M = \max\{|x_1|, |x_2|, \ldots, |x_{N-1}|\}$. Then, we have

$$T \subseteq B(0; 1 + M).$$

Thus, T is an infinite and bounded set in $\mathbb{R}$. In view of the Bolzano–Weierstrass Theorem, i.e., Theorem 4.2.10, T has an accumulation point p in $\mathbb{R}$.

Next, we next claim that $\{x_n\}$ converges to p. To this end, take any $\epsilon > 0$. There exists a positive integer N such that

$$|x_m - x_n| < \frac{\epsilon}{2} \quad \text{whenever } n \geq N.$$

Now, the ball $B_M\left(p; \frac{\epsilon}{2}\right)$ contains a point x_{n_0} with $n_0 \geq N$. Hence, if $n \geq N$, then we have

$$|x_n - p| \leq |x_n - x_{n_0}| + |x_{n_0} - p|$$
$$< \frac{\epsilon}{2} + \frac{\epsilon}{2}$$
$$= \epsilon,$$

i.e., $\lim_{n \to \infty} x_n = p$. This completes the proof.

Now, we can club Theorems 4.4.1 and 4.4.2 to have the following result.

Theorem 4.4.3. *In the Euclidean space* $\mathbb{R}$, *a sequence* $\{x_n\}$ *converges if and only if it is a Cauchy sequence.*

In the above context, note that Theorem 4.4.2 is very useful to prove the convergence of a sequence, when the limit is not known already. For example, in $\mathbb{R}$, consider the sequence $\{x_n\}$ defined by

$$x_n = 1 - \frac{1}{2} + \frac{1}{3} - \frac{1}{4} + \cdots + \frac{(-1)^{n-1}}{n} \quad \text{for every } n = 1, 2, \ldots.$$

We can prove that $\{x_n\}$ is a Cauchy sequence in $\mathbb{R}$ and so it converges to some limit (note that this limit is log 2!).

Definition 4.4.2. (1) A metric space (M, d) is said to be *complete* if every Cauchy sequence in M converges to a point in M;

(2) $S \subseteq M$ is said to be *complete* if the metric subspace (S, d) is complete.

We have already proved that $\mathbb{R}$ is complete and the Euclidean metric subspace $S = (0, 1]$ is not complete. It is worth answering the following question:

Is $\mathbb{Q}$, *the set of all rational numbers, with Euclidean metric complete, justifying the answer?*

It is also worthwhile to prove that $\mathbb{C} = \mathbb{R} \times \mathbb{R}$ (the set of all complex numbers), with the usual metric, is complete.

4.5   Exercises

Excercise 4.1 Prove that any open interval in $\mathbb{R}$ is an open set and any closed interval is a closed set.

Excercise 4.2 Find all the accumulation points of the following sets in $\mathbb{R}$. Find which of them are open or closed or neither:

(1) All integers;

(2) All rational numbers;

(3) the interval $(0, 1]$;

(4) the interval $(0, 1)$;

(5) All numbers $\frac{1}{n}$ for every $n = 1, 2, \ldots$;

(6) All numbers $\frac{1}{m} + \frac{1}{n}$ for every $m, n = 1, 2, \ldots$;

(7) All numbers $(-1)^n \left(1 + \frac{1}{n}\right)$ for every $n = 1, 2, \ldots$.

Excercise 4.3 Prove that any non-empty set S in $\mathbb{R}$ contains both rational and irrational numbers.

Excercise 4.4 Prove that the only sets in $\mathbb{R}$ which are both open and closed are ϕ and $\mathbb{R}$ itself.

Excercise 4.5 Prove that the interior of a set in $\mathbb{R}$ is open in $\mathbb{R}$.

Excercise 4.6 If $S \subseteq \mathbb{R}$, then prove that $int\, S$ is the longest open subset of S in the sense that $int\, S$ is the union of all subsets of $\mathbb{R}$ which are contained in S.

Excercise 4.7 If $S, T \subseteq \mathbb{R}$, then prove that

(1) $(int\, S) \bigcap (int\, T) = int\, (S \bigcap T)$;

(2) $(int\, S) \bigcup (int\, T) \subseteq int\, (S \bigcup T)$.

Excercise 4.8 Let $S, T \subseteq \mathbb{R}$. Prove that

(1) S' is closed in $\mathbb{R}$;

(2) If $S \subseteq T$, then $S' \subseteq T'$;

(3) $(S \cup T)' = S' \cup T'$;

(4) $(\overline{s})' = S'$.

Excercise 4.9 Prove that $\overline{s}$ is the smallest closed set containing S in the sense that $\overline{s}$ is the intersection of all closed subsets of $\mathbb{R}$ containing S.

Excercise 4.10 Let $S, T \subseteq \mathbb{R}$. Prove that $\overline{S \cap T} \subseteq \overline{S} \cap \overline{T}$ and, if S is open, then we have $S \cap \overline{T} \subseteq \overline{S \cap T}$.

Excercise 4.11 Prove Exercises 4.5–4.10 by replacing $\mathbb{R}$ by any metric spaces.

Excercise 4.12 Prove that any finite set in a metric space is closed.

Excercise 4.13 Let (M, d) be a metric space. For every $x, y \in M$, define a mapping $d' : M \times M \to \mathbb{R}$ by

$$d'(x, y) = \frac{d(x, y)}{1 + d(x, y)}.$$

Prove that d' is also a metric on M with the property:

$$0 \le d'(x, y) < 1.$$

(This exercise shows that if there is one metric defined on M, then we can generate any number of metrices on M).

Excercise 4.14 Let S be a subset of metric space M. Prove that

(1) $int\, S = M - \overline{M - S}$;

(2) $int\, (M - S) = M - \overline{S}$;

(3) $int\, (int\, A) = int\, A$.

Excercise 4.15 (1) If M is a metric space and $A_i \subseteq M$ for every $i = 1, 2, \ldots, n$, then prove that

$$int \left(\bigcap_{i=1}^{n} A_i \right) = \bigcap_{i=1}^{n} int\, A_i;$$

(2) If $\mathscr{F}$ is an infinite collection of subsets of M, then prove that

$$int \left(\bigcap_{A \in \mathscr{F}} A \right) \subseteq \bigcap_{A \in \mathscr{F}} (int\, A).$$

Give an example where equality does not hold.

Excercise 4.16 If $\mathscr{F}$ is a collection of subsets of a metric space M, then prove that

$$\bigcup_{A \in \mathscr{F}} (int\, A) \subseteq int \left(\bigcup_{A \in \mathscr{F}} A \right).$$

Give also an example of a finite collection $\mathscr{F}$, for which equality does not hold.

Excercise 4.17 In a metric space (M, d), if $\lim_{n \to \infty} x_n = x$ and $\lim_{n \to \infty} y_n = y$, then prove that

$$\lim_{n \to \infty} d(x_n, y_n) = d(x, y).$$

Excercise 4.18 Let A be a subset of a metric space (M, d). If A is complete, then prove that A is close. Prove that the converse also holds if M is complete.

Excercise 4.19 In $\mathbb{R}^2$, let $x = (x_1, x_2), y = (y_1, y_2) \in \mathbb{R} \times \mathbb{R}$. Define a mapping $d : \mathbb{R} \times \mathbb{R} \to \mathbb{R}$ by

$$d(x, y) = \max\{|x_1 - y_1|, |x_2 - y_2|\}.$$

Prove that d is a metric on $\mathbb{R}^2$ with respect to which $\mathbb{R}^2$ is complete.

Excercise 4.20 In $\mathbb{R}^2$, define a mapping $d : \mathbb{R} \times \mathbb{R} \to \mathbb{R}$ by

$$d(x, y) = |x_1 - y_1| + |x_2 - y_2|,$$

where $x = (x_1, x_2), y = (y_1, y_2) \in \mathbb{R} \times \mathbb{R}$. Prove that d is a metric on $\mathbb{R}^2$.

Excercise 4.21 Let $\{a_n\}$ be a sequence of real numbers such that

$$|a_{n+2} - a_{n+1}| \leq \frac{1}{2}|a_{n+1} - a_n| \quad \text{for every } n = 1, 2, \dots.$$

Prove that $\{a_n\}$ converges by showing that $\{a_n\}$ is a Cauchy sequence.

Excercise 4.22 (The Bolzano–Weierstrass Theorem) Prove that every bounded sequence of real numbers has a convergent subsequence.

5

Limits and Continuity

In this chapter, we first discuss the limit of a function, the right-hand limit, the left-hand limit, the infinite limits, the limits at infinity, the sequential definition of the limit, and Cauchy's Criterion for finite limits and present several results covering various properties of the limits with some examples. Next, we discuss monotonic, continuous and discontinuous functions by giving several related results with examples. Then, the notion of uniform continuity is discussed with results and examples. Finally, we give a brief discussion on the continuity and the uniform continuity in metric spaces. Several solved examples and chapter-end exercises are also included covering all topics discussed in the chapter.

5.1   The Limit of Functions

The concept of the limit is one of the basic concepts in mathematical analysis and is used to define the continuity, the derivatives, and the integrals.

Let $r > 0$ be a number and f be a function defined at all points in a neighborhood, $N(c, r) = \{x \in \mathbb{R} : 0 < |x - c| < 0\}$, of a point c except possibly at the point c itself.

Definition 5.1.1. A number l is said to be the *limit* of f at $x = c$ if, for every $\epsilon > 0$, there exists a number $\delta > 0$ such that

$$|f(x) - l| < \epsilon \ \text{ whenever } \ 0 < |x - c| < \delta,$$

i.e., $f(x) \in (l - \epsilon, l + \epsilon)$ for all $x \in (c - \delta, c + \delta)$ (except possibly c). We write as follows:

$$\lim_{x \to c} f(x) = l \ \text{ or } \ f(x) \to l \ \text{ as } \ x \to c.$$

Example 5.1.1. Using the definition of the limit, prove the following:

(1) $\lim\limits_{x \to c} \frac{x^2 - c^2}{x - c} = 2c$;

(2) $\lim\limits_{x \to 0} x^2 = 0$;

(3) $\lim\limits_{x \to 0} x \sin\left(\frac{1}{x}\right) = 0$;

(4) $\lim\limits_{x \to 0} x \cos\left(\frac{1}{x}\right) = 0$.

Solution. (1) Let $f(x) = \frac{x^2 - c^2}{x - c} = 2c$, where $x \neq c$. Let $\epsilon > 0$ be given. Then, we have

$$|f(x) - 2c| = \left|\frac{x^2 - c^2}{x - c} - 2c\right| = \left|\frac{(x - c)^2}{x - c}\right| = |x - c|.$$

If we choose $\delta = \epsilon$, then we have

$$|f(x) - 2c| < \epsilon \text{ whenever } 0 < |x - c| < \delta.$$

Hence, from the definition of the limit, we have

$$\lim_{x \to c} f(x) = 2c, \text{ i.e., } \lim_{x \to c} \frac{x^2 - c^2}{x - c} = 2c.$$

(2) Note that $|x^2 - 0| < \epsilon$ implies $|x| < \sqrt{\epsilon}$. Hence, for every $\epsilon > 0$, we can take $\delta = \sqrt{\epsilon}$ in the definition of the limit.

(3) Let $g(x) = x \sin\left(\frac{1}{x}\right)$, where $x \neq 0$. Let $\epsilon > 0$ be given. Then, we have

$$|g(x) - 0| = \left|x \sin\left(\frac{1}{x}\right)\right| = |x| \left|\sin\left(\frac{1}{x}\right)\right| \leq |x|$$

as $\left|\sin\left(\frac{1}{x}\right)\right| \leq 1$ for all $x \neq 0$. Thus, for every $\epsilon > 0$, we can take $\delta = \epsilon$, so that

$$|g(x) - 0| < \epsilon, \text{ i.e., } \left|x \sin\left(\frac{1}{x}\right) - 0\right| < \epsilon \text{ whenever } |x - 0| < \delta.$$

Hence, we have $\lim\limits_{x \to 0} x \sin\left(\frac{1}{x}\right) = 0$.

(4) Proceed as in (3).

As an immediate consequence of the definition of limit of a function, we have the following theorem.

Theorem 5.1.1. *If a function f possesses a finite limit at a point c, then a neighborhood $N(c, r)$ of c exists on which f is bounded.*

Theorem 5.1.2. *The limit of a function at a point, if it exists, is unique.*

Proof. Let $\lim\limits_{x \to c} = l$ and $\lim\limits_{x \to c} = m$. Suppose that $l \neq m$. Now, we show that this leads to a contradiction. In fact, let $\epsilon > 0$ be given. Since $\lim\limits_{x \to c} = l$, there exists $\delta_1 > 0$ such that

$$|f(x) - l| < \frac{\epsilon}{2} \text{ whenever } 0 < |x - c| < \delta_1.$$

Again, since $\lim_{x \to c} = m$, there exists $\delta_2 > 0$ such that

$$|f(x) - m| < \frac{\epsilon}{2} \quad \text{whenever} \quad 0 < |x - c| < \delta_2.$$

Now, for $0 < |x - c| < \min\{\delta_1, \delta_2\}$, we have

$$
\begin{aligned}
|l - m| &= |l - f(x) + f(x) - m| \\
&\leq |f(x) - l| + |f(x) - m| \\
&< \frac{\epsilon}{2} + \frac{\epsilon}{2} \\
&= \epsilon.
\end{aligned}
\tag{5.1}
$$

Since $\epsilon > 0$ is arbitrary, it follows from (5.1) that $|l - m| = 0$, that is, $l = m$, which is a contradiction. Hence, if $\lim_{x \to c}$ exists, then it is unique. This completes the proof.

5.2 Algebras of Limits

Theorem 5.2.1. *If* $\lim_{x \to c} f(x) = l$ *and* $\lim_{x \to c} g(x) = m$, *then we have the following:*

(1) $\lim_{x \to c} [f(x) \pm g(x)] = l \pm m$;

(2) $\lim_{x \to c} [f(x)g(x)] = lm$;

(3) $\lim_{x \to c} \left(\frac{f(x)}{g(x)} \right) = \frac{l}{m}$ *provided* $m \neq 0$.

Proof. (1) Let $\epsilon > 0$ be given. Since $\lim_{x \to c} = l$, there exists $\delta_1 > 0$ such that

$$|f(x) - l| < \frac{\epsilon}{2} \quad \text{whenever} \quad 0 < |x - c| < \delta_1. \tag{5.2}$$

Again, since $\lim_{x \to c} = m$, there exists $\delta_2 > 0$ such that

$$|f(x) - m| < \frac{\epsilon}{2} \quad \text{whenever} \quad 0 < |x - c| < \delta_2. \tag{5.3}$$

Choosing $\delta = \min\{\delta_1, \delta_2\}$, it follows from (5.2) and (5.3) that

$$
\begin{aligned}
|(f(x) \pm g(x)) - (l \pm m)| &= |(f(x) - l) \pm (g(x) - m)| \\
&\leq |f(x) - l| + |g(x) - m| \\
&< \frac{\epsilon}{2} + \frac{\epsilon}{2} \\
&= \epsilon
\end{aligned}
$$

whenever $0 < |x - a| < \delta$. Hence, we have $\lim_{x \to c}[f(x) \pm g(x)] = l \pm m$.

The result can evidently be extended to any finite number of functions.

(2) We have

$$|f(x)g(x) - lm| = |g(x)(f(x) - l) + l(g(x) - m)|$$
$$\leq |g(x)||f(x) - l| + |l||g(x) - m|.$$

Since $\lim_{x \to c} g(x) = m$, g is bounded in a neighborhood $N(c, r)$ of $x = c$. Hence, there exists $M > 0$ such that $|g(x)| < M$ whenever $0 < |x - c| < \delta'$ for some $\delta' > 0$. Then, we have

$$|f(x)g(x) - lm| < M|f(x) - l| + |l||g(x) - m|.$$

Let $\epsilon > 0$ be given. Since $\lim_{x \to c} f(x) = l$ and $\lim_{x \to c} g(x) = m$, we can find $\delta > 0$ with $\delta < \delta'$ such that

$$|f(x) - l| < \frac{\epsilon}{2M}, \quad |g(x) - m| < \frac{\epsilon}{2(|l| + 1)} \quad \text{whenever } 0 < |x - c| < \delta.$$

So, we have

$$|f(x)g(x) - lm| < M\frac{\epsilon}{2M} + |l|\frac{\epsilon}{2(|l| + 1)} < \epsilon \quad \text{whenever } 0 < |x - c| < \delta.$$

Hence, we have $\lim_{x \to c}[f(x)g(x)] = lm$.

The above result can evidently be extended to any finite number of functions.

(3) Let $\lim_{x \to c} f(x) = l$ and $\lim_{x \to c} g(x) = m \neq 0$. Then, the functions f and g are bounded in the neighborhood $N(c, r)$ of the point $x = c$. Hence, there exist $M > 0$ and $k > 0$ such that

$$|f(x)| < M \quad \text{whenever} \quad 0 < |x - c| < \delta_1$$

and

$$|g(x)| > k \quad \text{whenever} \quad 0 < |x - c| < \delta_2$$

for some $\delta_1 > 0$ and $\delta_2 > 0$. Thus, we can have

$$\left| \frac{f(x)}{g(x)} - \frac{l}{m} \right| = \left| \frac{f(x)}{g(x)} - \frac{f(x)}{m} + \frac{f(x)}{m} - \frac{l}{m} \right|$$
$$\leq \frac{|f(x)|}{|m||g(x)|}|g(x) - m| + \frac{1}{|m|}|f(x) - l|$$
$$\leq \frac{M}{|m|k}|g(x) - m| + \frac{1}{|m|}|f(x) - l| \tag{5.4}$$

whenever $0 < |x - c| < \delta_3 = \min\{\delta_1, \delta_2\}$. Since $\lim\limits_{x \to c} f(x) = l$ and $\lim\limits_{x \to c} g(x) = m \neq 0$, corresponding to any $\epsilon > 0$, we can find $\delta_4 > 0$ and $\delta_5 > 0$ such that

$$|f(x) - l| < \frac{|m|\epsilon}{2} \quad \text{whenever} \ \ 0 < |x - c| < \delta_4$$

and

$$|g(x) - m| < \frac{|m|k\epsilon}{2M} \quad \text{whenever} \ \ 0 < |x - c| < \delta_5.$$

Choosing $\delta = \min\{\delta_3, \delta_4, \delta_5\}$, it follows from (5.4) that

$$\left| \frac{f(x)}{g(x)} - \frac{l}{m} \right| < \epsilon \quad \text{whenever} \ \ 0 < |x - c| < \delta.$$

Hence, we have $\lim\limits_{x \to c} \left[\frac{f(x)}{g(x)} \right] = \frac{l}{m}$ provided $m \neq 0$. This completes the proof.

Theorem 5.2.2. (The Pinching or Sandwich Theorem) *Let f, g, and h defined on a deleted neighborhood U of a point c be such that $f(x) \leq g(x) \leq h(x)$ and $\lim\limits_{x \to c} f(x) = \lim\limits_{x \to c} h(x) = l$. Then, $\lim\limits_{x \to c} g(x) = l$.*

Proof. Let $\lim\limits_{x \to c} f(x) = \lim\limits_{x \to c} h(x) = l$. Then, corresponding to any $\epsilon > 0$, we can find a positive number δ such that

$$|f(x) - l| < \epsilon, \quad \text{i.e.,} \ \ l - \epsilon < f(x) < l + \epsilon$$

and

$$|h(x) - l| < \epsilon, \quad \text{i.e.,} \ \ l - \epsilon < h(x) < l + \epsilon$$

whenever $0 < |x - c| < \delta$. Since $f(x) \leq g(x) \leq h(x)$, we have

$$l - \epsilon < f(x) \leq g(x) \leq h(x) < l + \epsilon$$

whenever $0 < |x - c| < \delta$. So, we have

$$l - \epsilon < g(x) < l + \epsilon, \quad \text{i.e.,} \ |g(x) - l| < \epsilon$$

whenever $0 < |x-c| < \delta$. Hence, $\lim\limits_{x \to c} g(x)$ exists and equals to l. This completes the proof.

Example 5.2.1. Use the Sandwich Theorem to prove that

$$\lim_{x \to 0} x \cos \left(\frac{1}{x} \right) = 0.$$

Solution. We have

$$\left| x \cos \left(\frac{1}{x} \right) \right| \leq |x| \quad \text{for every} \ x \neq 0.$$

Hence, we have

$$-|x| \le x \cos\left(\frac{1}{x}\right) \le |x|.$$

Since $\lim_{x\to 0} -|x| = \lim_{x\to 0} |x| = 0$, it follows from the Sandwich Theorem that

$$\lim_{x\to 0} x \cos\left(\frac{1}{x}\right) = 0.$$

Theorem 5.2.3. *Let* $\lim_{x\to c} f(x) = 0$ *and* $g(x)$ *be bounded in some deleted neighborhood of* c. *Then, we have*

$$\lim_{x\to c} f(x)g(x) = 0.$$

Proof. Since the function $g(x)$ is bounded in some deleted neighborhood of c, there exist numbers $M > 0$ and $\delta_1 > 0$ such that

$$|g(x)| \le M \quad \text{whenever} \quad 0 < |x - c| < \delta_1. \tag{5.5}$$

Since $\lim_{x\to c} f(x) = 0$, corresponding to any $\epsilon > 0$, there exists $\delta_2 > 0$ such that

$$|f(x)| = |f(x) - 0| < \frac{\epsilon}{M} \quad \text{whenever} \quad 0 < |x - c| < \delta_2. \tag{5.6}$$

Then, for $0 < |x - c| < \min\{\delta_1, \delta_2\}$, it follows from (5.5) and (5.6) that

$$|f(x)g(x) - 0| = |f(x)g(x)| = |f(x)||g(x)| < \frac{\epsilon}{M} M = \epsilon.$$

Hence, we have

$$\lim_{x\to c} f(x)g(x) = 0.$$

This completes the proof.

Example 5.2.2. Show that $\lim_{x\to 0} x \sin\left(\frac{1}{x}\right) = 0$.

Solution. Here $\lim_{x\to 0} x = 0$ and $\left|\sin\left(\frac{1}{x}\right)\right| \le 1$ for all $x \ne 0$, i.e., $\sin\left(\frac{1}{x}\right)$ is bounded in a deleted neighborhood of 0. Then, we have

$$\lim_{x\to 0} x \sin\left(\frac{1}{x}\right) = 0$$

by taking $f(x) = x$ and $g(x) = \sin\left(\frac{1}{x}\right)$ in Theorem 5.2.3 for $c = 0$.

5.3   Right-Hand and Left-Hand Limits

If x approaches c from the right (or from above), that is, from values of x larger than c, then the limit of f as defined before is called the *right hand limit* of $f(x)$, which is written as follows:

$$\lim_{x\to c+} f(x) \quad \text{or} \quad f(c+) \quad \text{or} \quad \lim_{x\to c+0} f(x) \quad \text{or} \quad f(c+0).$$

Definition 5.3.1. (The Right-Hand Limit) We say that

$$\lim_{x\to c+} f(x) = l$$

if, for every $\epsilon > 0$, there exists a number $\delta > 0$ such that

$$|f(x) - l| < \epsilon \text{ whenever } 0 < x - c < \delta \text{ (or } 0 < x < c + \delta).$$

If x approaches c from the left (or from below), that is, from values of x smaller than c, the limit of f is called the *left-hand limit* and is written as follows:

$$\lim_{x\to c-} f(x) \quad \text{or} \quad f(c-) \quad \text{or} \quad \lim_{x\to c-0} f(x) \quad \text{or} \quad f(c-0).$$

Definition 5.3.2. (The Left-Hand Limit) We say that

$$\lim_{x\to c-} f(x) = l$$

if, for every $\epsilon > 0$, there exists a number $\delta > 0$ such that

$$|f(x) - l| < \epsilon \text{ whenever } \delta < x - c < 0 \text{ (or } c - \delta < x < c).$$

If both right-hand and left-hand limits of f, as x approaches c, exist and are equal in value, then their common value, evidently, will be the limit of f as x approaches c. However, if either or both of these two limits do not exist, then the limit of f as x approaches c does not exist.

Even if both right-hand and left-hand limits exist, but are not equal in value, then also the limit of f as x approaches c does not exist.

In fact, we have the following theorem:

Theorem 5.3.1. *Let f be defined on a deleted neighborhood of 'c'. Then, the following are equivalent:*

(1) $\lim_{x\to c} f(x)$ *exists and equals l;*

(2) *Both $\lim_{x\to c+} f(x)$ and $\lim_{x\to c-} f(x)$ exist and are equal to l.*

Proof. (1) $\implies$ (2) Let $\lim_{x\to c} f(x) = l$. Then, for every $\epsilon > 0$, there exists a number $\delta > 0$ such that

$$|f(x) - l| < \epsilon \text{ whenever } c - \delta < x < c + \delta \text{ with } x \neq c.$$

It follows that

$$|f(x) - l| < \epsilon \text{ whenever } c - \delta < x < c \tag{5.7}$$

and

$$|f(x) - l| < \epsilon \text{ whenever } c < x < c + \delta. \tag{5.8}$$

Hence, $\lim\limits_{x \to c-} f(x)$ and $\lim\limits_{x \to c+} f(x)$ both exist and are equal to l by (5.7) and (5.8), respectively.

(2) $\implies$ (1) Let $\lim\limits_{x \to c+} f(x) = \lim\limits_{x \to c-} f(x) = l$. Then, for every $\epsilon > 0$, there exist $\delta_1 > 0$ and $\delta_2 > 0$ such that

$$|f(x) - l| < \epsilon \quad \text{whenever} \quad c < x < c + \delta_1 \tag{5.9}$$

and

$$|f(x) - l| < \epsilon \quad \text{whenever} \quad c - \delta_2 < x < c. \tag{5.10}$$

If we choose $\delta = \min\{\delta_1, \delta_2\}$, then $c + \delta \le c + \delta_1$ and $c - \delta_2 \le c - \delta$. So, it follows from (5.9) and (5.10) that

$$|f(x) - l| < \epsilon \quad \text{whenever} \quad c - \delta < x < c + \delta \quad \text{with} \quad x \ne c,$$

i.e.,

$$|f(x) - l| < \epsilon \quad \text{whenever} \quad 0 < |x - c| < \delta.$$

Hence, it follows that $\lim\limits_{x \to c} f(x)$ exists and equals to l. This completes the proof.

Example 5.3.1. Prove that

$$\lim_{x \to 0+} \sqrt{x} = 0.$$

Solution. Let $\epsilon > 0$ be given. If we choose $\delta = \epsilon^2$, then we have

$$|\sqrt{x} - 0| < \epsilon \quad \text{whenever} \quad 0 < x - 0 < \delta.$$

So, by the definition of the right-hand limit, we have

$$\lim_{x \to 0+} \sqrt{x} = 0.$$

Example 5.3.2. Evaluate

$$\lim_{x \to 0} \frac{2x - |x|}{x}.$$

Solution. We have

$$\lim_{x \to 0+} \frac{2x - |x|}{x} = \lim_{x \to 0+} \frac{2x - x}{x} = 1$$

and

$$\lim_{x \to 0-} \frac{2x - |x|}{x} = \lim_{x \to 0-} \frac{2x - (-x)}{x} = 3.$$

Thus, we have

$$\lim_{x \to 0+} \frac{2x - |x|}{x} \neq \lim_{x \to 0-} \frac{2x - |x|}{x}.$$

So, $\lim\limits_{x \to 0} \frac{2x-|x|}{x}$ does not exist.

5.4   Infinite Limits and Limits at Infinity

Definition 5.4.1. Let $f(x)$ be a function defined for all points in some neighborhood of a point $x = c$ except possibly at $x = c$ itself. Then, we say that

$$\lim_{x \to c} f(x) = \infty$$

if, for every $K > 0$, there exists a number $\delta > 0$ such that

$$f(x) > K \quad \text{whenever} \quad 0 < |x - c| < \delta.$$

Definition 5.4.2. Let $f(x)$ be a function defined for all points in some neighborhood of a point $x = c$ except possibly at $x = c$ itself. Then, we say that

$$\lim_{x \to c} f(x) = -\infty$$

if, for every $L < 0$, there exists a number $\delta > 0$ such that

$$f(x) < L \quad \text{whenever} \quad 0 < |x - c| < \delta.$$

Definition 5.4.3. Let $f(x)$ be a function defined on $x > M$ for some M. Then, we say that

$$\lim_{x \to \infty} f(x) = l$$

if, for every $\epsilon > 0$, there exists a number $K > 0$ such that

$$|f(x) - l| < \epsilon \quad \text{whenever} \quad x > K.$$

Definition 5.4.4. Let $f(x)$ be a function defined on $x < M$ for some M. Then, we say that

$$\lim_{x \to -\infty} f(x) = l$$

if, for every $\epsilon > 0$, there exists a number $L < 0$ such that

$$|f(x) - l| < \epsilon \quad \text{whenever} \quad x < L.$$

Let us now look at limits at infinity that are also infinite in value. There are four such possible limits. We define one of them, and the other three can be defined similarly.

Definition 5.4.5. Let $f(x)$ be a function defined on $x > M$ for some M. Then, we say that

$$\lim_{x \to \infty} f(x) = \infty$$

if, for every $L > 0$, there exists a number $K > 0$ such that

$$f(x) > L \quad \text{whenever} \quad x > K.$$

Example 5.4.1. Prove that

$$\lim_{x \to -\infty} \frac{1}{x} = 0.$$

Solution. Let $\epsilon > 0$ be any number. We need to find a number $K < 0$ such that

$$\left| \frac{1}{x} - 0 \right| < \epsilon \quad \text{whenever} \quad x < K.$$

Now, $x < -\frac{1}{\epsilon}$ implies $\frac{1}{|x|} < \epsilon$. Thus, if we choose $K = -\frac{1}{\epsilon} < 0$, then we have

$$\left| \frac{1}{x} - 0 \right| = \frac{1}{|x|} < \epsilon \quad \text{whenever} \quad x < K.$$

Hence, we have

$$\lim_{x \to -\infty} \frac{1}{x} = 0.$$

5.5 Certain Important Limits

The readers should be familiar with the following limits:

(1) $\lim\limits_{x \to \infty} \left(1 + \frac{1}{x} \right)^x = e, \quad \lim\limits_{x \to \infty} \left(1 - \frac{1}{x} \right)^{-x} = e, \quad \lim\limits_{x \to 0} (1 + x)^{\frac{1}{x}} = e;$

(2) $\lim\limits_{x \to 0} \frac{\log(1+x)}{x} = 1;$

(3) $\lim\limits_{x \to 0} \frac{a^x - 1}{x} = \log a$ for every $a > 0;$

(4) $\lim\limits_{x \to c} \frac{x^n - c^n}{x - c} = nc^{n-1}$ for every $n \neq 0$ and $c \neq 0$ if $n = 0;$

(5) $\lim\limits_{x \to 0} \frac{\sin x}{x} = 1, \quad \lim\limits_{x \to 0} \cos x = 1.$

5.6 Sequential Definition of Limit of a Function

Let f be a function defined for all points in some neighborhood U of a point c except possibly at the point c itself. Then, the following is an alternative definition of the limit of such functions in terms of limits of sequences known as the sequential definition of the limit of functions.

Definition 5.6.1. A number l is called the *limit* of f as x tends to c if the limit of the sequence $\{f(x_n)\}$ exists and is equal to l for every sequence $\{x_n\}$, $x_n \in U - \{c\}$ for every $n = 1, 2, \ldots$, convergent to c.

Both the definitions of the limit of a function, i.e., Definitions 5.1.1 and 5.6.1 are equivalent. In fact, we have the following result.

Theorem 5.6.1. *Let $c \in \mathbb{R}$, U be a neighborhood of c and f be a function defined everywhere on U except possibly at c. Then, the following are equivalent:*

(1) $\lim_{x \to c} f(x) = l$;

(2) $f(x_n) \to l$ as $n \to \infty$ for every sequence $\{x_n\}$, $x_n \in U \setminus \{c\}$ for every $n = 1, 2, \ldots$, which converges to c.

Proof. (1) $\Longrightarrow$ (2) $\lim_{x \to c} f(x) = l$ exists in the sense of Definition 5.1.1. Then, corresponding to every $\epsilon > 0$, there exists $\delta > 0$ such that

$$|f(x) - l| < \epsilon \quad \text{whenever} \quad 0 < |x - c| < \delta. \tag{5.11}$$

Also, suppose that there exists a sequence $\{x_n\}$, $x_n \neq c$ for every 1, 2, $\ldots$, converging to c. Then, there exists $m \in \mathbb{N}$ such that

$$|x_n - c| < \delta \quad \text{for every} \quad n \geq m. \tag{5.12}$$

Hence, it follows from (5.11) and (5.12) that

$$|f(x_n) - l| < \epsilon \quad \text{for every} \quad n \geq m.$$

Therefore, the sequence $\{f(x_n)\}$ converges to l for every sequence $\{x_n\}$, $x_n \neq c$ for every $n = 1, 2, \ldots$, which converges to c proving that $\lim_{x \to c} f(x) = l$ exists in the sense of Definition 5.6.1 too.

(2) $\Longrightarrow$ (1) Suppose that $\lim_{x \to c} f(x) = l$ exists in the sense of Definition 5.6.1 and this limit does not exist for f in the sense of Definition 5.1.1. Then, there must exist at least one ϵ, say ϵ_1 for which there is no δ in the sense of Definition 5.1.1, i.e., for every $\delta > 0$, there is $x = x^{(\delta)}$ satisfying

$$0 < |x - c| < \delta, \quad |f(x^{(\delta)}) - l| \geq \epsilon_1.$$

Let us choose $\delta = 1, \frac{1}{2}, \frac{1}{3}, \ldots$ successively. Then, corresponding to each value of δ, we must have an x_n such that

$$|x_n - c| < \frac{1}{n}(x_n \neq c)$$

and

$$|f(x_n) - l| \geq \epsilon_1 \text{ for every } n = 1, 2, \ldots.$$

The above two relations show that $\{x_n\}$ $(x_n \neq c)$ converges to c while $\{f(x_n)\}$ does not converge to l, which is a contradiction to our supposition. Hence, if $\lim_{x \to c} f(x) = l$ exists in the sense of Definition 5.6.1, then this limit must exist for f in the sense of Definition 5.1.1. This completes the proof.

Example 5.6.1. Let

$$f(x) = \begin{cases} 1, & \text{if } x \in \mathbb{Q}, \\ -1, & \text{if } x \notin \mathbb{Q}. \end{cases}$$

Show that $\lim_{x \to c} f(x)$ does not exist for every real number c.

Solution. Let $\{x_n\}$ be a sequence of rational numbers such that $x_n \to c \, (x_n \neq c)$ as $n \to \infty$. Let $\{y_n\}$ be another sequence of irrational numbers such that $y_n \to c \, (y_n \neq c)$ as $n \to \infty$. Then, we have

$$f(x_n) = 1, \quad f(y_n) = -1 \text{ for every } n = 1, 2, \ldots.$$

Hence, we have

$$f(x_n) \to 1 \text{ as } n \to \infty \text{ whereas } f(y_n) \to 1 \text{ as } n \to \infty.$$

Therefore, $\lim_{x \to c} f(x)$ does not exist for every real number c in the sense of Definition 5.6.1.

5.7 Cauchy's Criterion for Finite Limits

Theorem 5.7.1. *The following are equivalent:*

    (1) *A function f tends to a finite limit as x tends to c;*

    (2) *For every $\epsilon > 0$, there exists a neighborhood U of c such that*

$$|f(x) - f(y)| < \epsilon \text{ for every } x, y \in U, x, y \neq c.$$

Proof. (1) $\implies$ (2) Let $\lim_{x \to c} f(x) = l$ be a finite number. So, corresponding to every $\epsilon > 0$, there exists a deleted neighborhood U of c such that

$$|f(x) - l| < \frac{\epsilon}{2} \text{ whenever } x \in U. \tag{5.13}$$

Let $x, y \in U$. Then, from (5.13), we have

$$|f(x) - l| < \frac{\epsilon}{2}, \quad |f(y) - l| < \frac{\epsilon}{2}$$

and so

$$|f(x) - f(y)| \leq |f(x) - l| + |f(y) - l| < \frac{\epsilon}{2} + \frac{\epsilon}{2} = \epsilon.$$

(2) $\implies$ (1) Suppose that, for every $\epsilon > 0$, there exists a deleted neighborhood U of c such that

$$|f(x) - f(y)| < \epsilon \text{ for every } x, y \in U. \tag{5.14}$$

Let $\{x_n\}$, $x_n \neq c$ for every $n = 1, 2, \ldots$, be an arbitrary sequence tending to c such that there exists $p \in \mathbb{N}$ such that $x_n, x_m \in U$ for every $n, n \geq p$. Then, it follows from (5.14) that

$$|f(x_n) - f(x_m)| < \epsilon \text{ for every } n, m \geq p.$$

So, by Cauchy's General Principle of the convergence, the sequence $\{f(x_n)\}$ tends to a limit. Thus, we have proved that $\lim f(x_n)$ exists for every sequence $\{x_n\}$ $(x_n \neq c)$ converging to c.

Now, we establish that the sequence $\{f(x_n)\}$ must tend to the same limit corresponding to all possible different sequences $\{x_n\}$ tending to c. In fact, suppose, if possible, the sequences $\{f(x_n)\}$ and $\{f(y_n)\}$ tend to l_1 and l_2, respectively, corresponding to two sequences $\{x_n\}$ $(x_n \neq c)$ and $\{y_n\}$ $(y_n \neq c)$ tending to c. Construct the sequence $\{x_1, y_1, x_2, y_2, \ldots\}$ which converges to c. So, as before, the sequence $\{f(x_1), f(y_1), f(x_2), f(y_2), \ldots\}$ converges, which is possible only if $l_1 = l_2$. This completes the proof.

We can have similar theorem for the existence of finite limit of a function at infinity (see Example 5.12).

5.8 Monotonic Functions

The monotonicity of a function tells us if the function is increasing or decreasing. A function is increasing when its graph rises from left to right. A function is decreasing when its graph falls from left to right. So, the function $f(x) = \sin x$ is increasing on the interval $[0, \frac{\pi}{2}]$ and decreasing on $[\frac{\pi}{2}, \pi]$. Now, let us think of a function f which first falls, then rises, and then falls again. So, it is not decreasing and not increasing, but it is neither non-decreasing nor non-increasing.

Now, we begin with the concept of increasing and decreasing functions for a single point of the domain of the definition.

Definition 5.8.1. (1) A function f is said to be *increasing* at a point $x = c$ if, for every x in the neighborhood $(c - \delta, c + \delta)$, $\delta > 0$ of c,

$$f(c - \delta) \leq f(x) \leq f(c + \delta);$$

(2) A function f is said to be *decreasing* at $x = c$ if, for every x in the neighborhood $(c - \delta, c + \delta)$, $\delta > 0$ of c,

$$f(c - \delta) \geq f(x) \geq f(c + \delta);$$

(3) A function f defined on a subset of $\mathbb{R}$ is said to be *monotonic* if it is either entirely non-increasing or entirely non-decreasing.

Definition 5.8.2. (1) A function f defined on a subset of $\mathbb{R}$ is said to be *monotonically increasing* (also *increasing* or *non-decreasing*) if, for every $x, y \in \mathbb{R}$ such that $x < y$, we have $f(x) \leq f(y)$;

(2) A function f is said to be *monotonically decreasing* (also *decreasing* or *non-increasing*) if, for every $x, y \in \mathbb{R}$ such that $x < y$, we have $f(x) \geq f(y)$.

Thus, a monotonically increasing function preserves the order, and a monotonically decreasing function reverses the order. A function is termed as strictly increasing or strictly decreasing if the strict inequality holds in the above definition, i.e.,

(3) A function f is said to be *strictly increasing* if, whenever $x < y$, $f(x) < f(y)$;

(4) A function f is said to be *strictly decreasing* if, whenever $x < y$, $f(x) > f(y)$.

If f is a strictly monotonic function, then f is one-to-one on its domain (because, for $x \neq y$, either $x < y$ or $x > y$ and so, by the monotonicity, either $f(x) < f(y)$ or $f(x) > f(y)$ and so $f(x) \neq f(y)$) and, if R is the range of f, then there is an inverse function on R for f.

In order to stress to include the possibility of repeating the same value at successive arguments, the terms "weakly increasing" and "weakly decreasing" may be used instead of "increasing" and "decreasing", respectively.

A function is said to be *unimodal* if it is monotonically increasing up to some point (the mode) and then monotonically decreasing.

Example 5.8.1. Prove that the function $f(x) = x^2 + 1$ is strictly increasing for every $x \geq 0$.

Solution. Let us take two arbitrary points $x_1 \geq 0$ and $x_2 \geq 0$ such that

$$x_1 < x_2.$$

Now, we have

$$f(x_1) - f(x_2) = (x_1^2 + 1) - (x_2^2 + 1) = x_1^2 - x_2^2 = (x_1 + x_2)(x_1 - x_2).$$

Since $x_1 + x_2 > 0$ and $x_1 - x_2 < 0$, we have $f(x_1) - f(x_2) < 0$, i.e.,

$$f(x_1) < f(x_2).$$

5.9   The Four Functional Limits at a Point

Let f be a function defined on (a, b), and let $a < c < b$. Let $L(h)$ and $l(h)$ denote the supremum and infimum of f in the right-hand neighborhood $(c, c + h)$ of c for some $h > 0$, and assign a sequence $\{h_n\}$ to h of diminishing values $h_1, h_2, \cdots$ such that $h_n \to 0$ as $n \to \infty$.

Then, $\{L(h_n)\}$ is non-increasing and so possesses a lower limit. Also, $\{l(h_n)\}$ is non-decreasing and so possesses an upper limit. These limits are called the *upper limit* and *lower limit* of f at the point c on the right, respectively; they are denoted by the symbols $\overline{f(c + 0)}$ and $\underline{f(c + 0)}$, respectively. Similarly, if a left-hand neighborhood $(c - h, c)$ of c for some $h > 0$ is considered, then the *lower limit* of $L(h)$ and the *upper limit* of $l(h)$ are defined to be the upper and lower limits of f at c on the left, respectively, and they are denoted by the symbols $\overline{f(c - 0)}$ and $\underline{f(c - 0)}$, respectively.

The numbers $\overline{f(c + 0)}$, $\underline{f(c + 0)}$, $\overline{f(c - 0)}$, and $\underline{f(c - 0)}$ are called the *four functional limits* of f at c.

If $\overline{f(c + 0)} = \underline{f(c + 0)}$, then their common value is said to be the *limit* of f at the point c on the right, which is denoted by $f(c + 0)$. Similarly, if $\overline{f(c - 0)} = \underline{f(c - 0)}$, then their common value is said to be the *limit* of f at c on the left, which is denoted by $f(c - 0)$.

If all the four limits are equal, then their common value is the *limit* of f at the point c.

5.10   Continuous and Discontinuous Functions

If the graph of a function $f : \mathbb{R} \to \mathbb{R}$ in the Cartesian plane is a continuous curve (i.e., a curve without breaks or holes), then it is natural to call such a function f continuous. However, not all functions are easy to draw.

Now, we need to use the definition of the continuity to determine a function's continuity. A continuous function with a continuous inverse function is called a *homeomorphism*. There are several different definitions of continuity of a function.

Let $f : S \to \mathbb{R}$ be function defined on a subset S of the set $\mathbb{R}$. Some possible choices of S (the domain of f) include $S = \mathbb{R}$ or $S = [a, b], a, b \in \mathbb{R}$ (i.e., S is a closed interval) or $S = (a, b)$ for every $a, b \in \mathbb{R}$ (i.e., S is an open interval).

In the next definition, we assume that S (the domain of f) does not have any isolated points. For example, if $S = [a, b]$ or (a, b) for every $a, b \in \mathbb{R}$, then S has no isolated points.

Definition 5.10.1. The function f is said to be *continuous* at a point $c \in S$ if the $\lim f(x)$ exists, as $x \to c$ through S, and is equal to $f(c)$, i.e.,

$$\lim_{x \to c} f(x) = f(c).$$

This means the following three conditions:

(1) f has to be defined at the point c a guaranteed by the requirement that c is in the domain of f;

(2) the limit on the left-hand side of the above equation must have to exist;

(3) the value of this limit must be equal to $f(c)$.

The following definition is popularly known as ϵ–δ (epsilon–delta) definition of continuous functions.

Definition 5.10.2. The function f is said to be *continuous* at the point $x_0 \in S$ if, for every $\epsilon > 0$, however small, there exists $\delta > 0$ such that, for every $x \in S$,

$$|f(x) - f(x_0)| < \epsilon \text{ whenever } |x - x_0| < \delta.$$

Now, we give a more general definition of the continuity in terms of neighborhoods, and this definition guarantees that a function is automatically continuous at every isolated point of its domain.

Definition 5.10.3. (1) A function f is said to be *continuous* at a point c of its domain if, for every neighborhood V of $f(c)$, there is a neighborhood U of c such that $f(x) \in V$ whenever $x \in U$;

(2) A function f is said to be *continuous* as a whole if it is continuous at every point of its domain.

Theorem 5.10.1. *The function f is continuous at $c \in S$ if and only if, for any sequence $\{x_n\} \subset S$ which converges to c, the corresponding sequence $\{f(x_n)\}$ converges to $f(c)$, i.e., $\lim_{n \to \infty} f(x_n) = f(c)$ whenever $\{x_n\} \subset S$ with $\lim_{n \to \infty} x_n = c$.*

Proof. Let the function f be continuous at $c \in S$ and $\{x_n\} \subset S$ be convergent to c. Since f is continuous at c, corresponding to every $\epsilon > 0$, there exists $\delta > 0$ such that

$$|f(x) - f(c)| < \epsilon \text{ whenever } 0 < |x - c| < \delta. \tag{5.15}$$

Also, since $\lim_{n \to \infty} x_n = c$, there exists $m \in \mathbb{N}$ such that

$$|x_n - c| < \delta \text{ for every } n \geq m. \tag{5.16}$$

Putting $x = x_n$ in (5.15) and then using (5.16), we get

$$|f(x_n) - f(c)| < \delta \text{ for every } n \geq m$$

and so

$$\lim_{n \to \infty} f(x_n) = f(c).$$

Next, suppose that f is not continuous at $c \in S$. Then, there exists $\epsilon > 0$ such that, for every $\delta > 0$, there exists $x \in S$ that

$$|x - c| < \delta, \text{ but } |f(x) - f(c)| \geq \epsilon.$$

By choosing $\delta = \frac{1}{n}$, we can find $x_n \in S$ for every $n \in \mathbb{N}$ such that

$$|x_n - c| < \frac{1}{n}, \text{ but } |f(x_n) - f(c)| \geq \epsilon.$$

Thus, $\{x_n\} \subset S$ converges to c, but $\{f(x_n)\}$ does not converge to $f(c)$. Hence, if $\lim_{n \to \infty} f(x_n) = f(c)$ whenever $\{x_n\} \subset S$ with $\lim_{n \to \infty} x_n = c$ for some $c \in S$, then f is continuous at c. This completes the proof.

Example 5.10.1. Prove that the function $f(x) = x^2$ is continuous at $x = 5$.

Solution. Clearly, we have $\lim_{x \to 5} f(x) = 5^2 = 25$ and $f(5) = 25$. So, we have

$$\lim_{x \to 5} f(x) = f(5).$$

Hence, f is continuous at $x = 5$.

Example 5.10.2. Examine the continuity of the following function:

$$f(x) = \begin{cases} \frac{x^5 e^{\frac{1}{x}}}{1 + e^{\frac{1}{x}}}, & \text{if } x \neq 0, \\ 0, & \text{if } x = 0, \end{cases}$$

at $x = 0$.

Solution. We have

$$\lim_{x \to 0-} f(x) = \lim_{x \to 0-} \frac{x^5 e^{\frac{1}{x}}}{1 + e^{\frac{1}{x}}} = 0$$

and

$$\lim_{x \to 0+} f(x) = \lim_{x \to 0+} \frac{x^5}{e^{-\frac{1}{x}} + 1} = 0.$$

So, we have

$$\lim_{x \to 0-} f(x) = \lim_{x \to 0+} f(x) = \lim_{x \to 0} f(x) = 0.$$

Also, we have

$$\lim_{x \to 0} f(x) = f(0).$$

Hence, the function f is continuous at 0.

Example 5.10.3. Examine the continuity of the function $f(x) = \frac{x+|x|}{2x}$.

Solution. If $x > 0$, then $f(x) = \frac{x+x}{2x} = 1$. If $x < 0$, then $f(x) = \frac{x-x}{2x} = 0$. So, if $x > 0$ or $x < 0$, the function f is continuous. However, f is not continuous at $x = 0$ as it is not defined at $x = 0$.

Example 5.10.4. Examine the continuity of the following function:

$$f(x) = \begin{cases} x, & \text{if } x \notin \mathbb{Q}, \\ -x, & \text{if } x \in \mathbb{Q}. \end{cases}$$

Solution. Let $c \neq 0$ be any rational number. Then, for every $n = 1, 2, \ldots$, we can choose an irrational number c_n such that

$$|c_n - c| < \frac{1}{n}.$$

Thus, we have

$$\lim_{n \to \infty} c_n = c.$$

Now, we have

$$f(c) = -c$$

and

$$\lim_{n \to \infty} f(c_n) = \lim_{n \to \infty} c_n = c.$$

Thus, we have

$$\lim_{n \to \infty} f(c_n) \neq f(c), \ c \neq 0.$$

Hence, by Theorem 5.10.1, the function f is not continuous at any non-zero rational number. Similarly, the function f can be shown to be not continuous at every irrational number.

Finally, consider the point $x = 0$. Let $\epsilon > 0$ be given. Then, we can choose $\delta = \epsilon$ such that $|x| = |x - 0| < \delta$ implies

$$|f(x) - f(0)| = |-x| = |x| < \epsilon$$

if x is rational and $|x| = |x - 0| < \delta$ implies

$$|f(x) - f(0)| = |x| < \epsilon$$

if x is irrational. Thus, for every $\epsilon > 0$, there exists $\delta > 0$ such that

$$|f(x) - f(0)| < \epsilon \text{ whenever } |x - 0| < \delta.$$

Hence, the function f is continuous at $x = 0$.

A function f is said to be *discontinuous* at a point c of its domain when it is not continuous at a point c. Then, the point c is called as the *point of discontinuity* of the function f. Discontinuities can be classified as removable discontinuity, discontinuity of the first kind (or ordinary discontinuity or jump discontinuity), discontinuity of the second kind, mixed discontinuity, or infinite discontinuity.

Definition 5.10.4. (The Removable Discontinuity) A function f is said to have a *removable discontinuity* at a point c if $\lim_{x \to c} f(x)$ exists, but is not equal to $f(c)$ (which may or may not exist), i.e., if $f(c+0) = f(c-0) \neq f(c)$. Such a discontinuity can be fixed by redefining the function.

Example 5.10.5. The function $f(x) = \frac{x^2 - 3x}{x^2 - 9}$ has a removable discontinuity at $x = 3$. We can redefine it as follows:

$$f(x) = \begin{cases} \frac{x^2 - 3x}{x^2 - 9} & \text{for every } x \neq 3, \\ \frac{1}{2} & \text{for } x = 3, \end{cases}$$

so that it becomes a continuous function.

Definition 5.10.5. (The Discontinuity of the First Kind or Ordinary Discontinuity or Jump Discontinuity)

(1) A function f is said to have a *discontinuity of the first kind* at c if both $f(c+0)$ and $f(c-0)$ exist but are not equal;

(2) The point c is said to be a *point of discontinuity of the first kind* from the left or right according as

$$f(c-0) \neq f(c) = f(c+0) \quad \text{or} \quad f(c-0) = f(c) \neq f(c+0).$$

Example 5.10.6. The following function:

$$f(x) = \begin{cases} x - 4, & \text{if } x < 4, \\ x + 4, & \text{if } x \geq 4 \end{cases}$$

has a jump discontinuity at $x = 4$.

Definition 5.10.6. (The Discontinuity of the Second Kind)

(1) A function f is said to have a *discontinuity of the second kind* at c if none of the limits $f(c+0)$ and $f(c-0)$ exists;

(2) The point c is said to be a *point of discontinuity of the second kind* from the left or right according as $f(c-0)$ or $f(c+0)$ does not exist.

Example 5.10.7. The function $f(x) = \sqrt{x+2}$ has a discontinuity of the second kind from the left at $x = -2$ because $f(-2-0)$ does not exist as the function is not defined when $x < -2$.

Definition 5.10.7. (The Mixed Discontinuity) A function f is said to have a *mixed discontinuity* at c if f has a discontinuity of the second kind on one side of c and on the other side a discontinuity of the first kind or may be continuous.

Example 5.10.8. The following function:

$$f(x) = \begin{cases} \sin \frac{1}{x}, & \text{if } x > 0, \\ 0, & \text{if } x \leq 0 \end{cases}$$

has a mixed discontinuity at $x = 0$ because f is continuous from the left of $x = 0$ but has discontinuity of the second kind from the right of $x = 0$.

Definition 5.10.8. (The Infinite Discontinuity) A function f is said to have an *infinite discontinuity* at a point c if one or more of the functional limits $\overline{f(c+0)}$, $f(c+0)$, $\overline{f(c-0)}$, and $f(c-0)$ are $+\infty$ or $-\infty$ and f is discontinuous at c.

Infinite discontinuities are sometimes referred to as *essential discontinuities* as such points of discontinuity are considered to be "more severe" than either removable or jump discontinuities.

It is clear from the above definition that if f is discontinuous at c and is unbounded in every neighborhood of c, then f must have an infinite discontinuity at c.

Example 5.10.9. The function $f(x) = -\frac{1}{x^2}$ has as infinite discontinuity at $x = 0$.

5.11   Some Theorems on the Continuity

Theorem 5.11.1. (1) *If f and g are two continuous functions at a point c, then $f \pm g$ is continuous at c;*

(2) *If f and g are continuous at a point c, then fg is continuous at c;*

(3) *If f and g are continuous at a point c and $g(c) \neq 0$, then $\frac{f}{g}$ is continuous at c.*

Proof. (1) Let f and g be continuous at c. So, we have

$$\lim_{x \to c} f(x) = f(c), \quad \lim_{x \to c} g(x) = g(c).$$

Now, we have

$$\lim_{x \to c} (f \pm g)(x) = \lim_{x \to c} \{f(x) \pm g(x)\}$$
$$= \lim_{x \to c} f(x) \pm \lim_{x \to c} g(x)$$
$$= f(a) \pm g(c) = f \pm g(c).$$

Hence, $f \pm g$ is continuous at c.

Similarly, using the properties of the limit, we can prove conclusions (2) and (3). This completes the proof.

Theorem 5.11.2. *The following are equivalent:*

(1) *A function f is continuous at c;*

(2) *For every $\epsilon > 0$, there exists $\delta > 0$ such that*

$$|f(x_1) - f(x_2)| < \epsilon \text{ whenever } x_1, x_2 \in (c - \delta, c + \delta).$$

Proof. (1) $\implies$ (2) Let f be continuous at c. Then, for every $\epsilon > 0$, there exists $\delta > 0$ such that

$$|f(x) - f(c)| < \frac{1}{2}\epsilon \text{ whenever } |x - c| < \delta.$$

So, we have

$$|f(x_1) - f(x_2)| = |f(x_1) - f(c) + f(c) - f(x_2)|$$
$$\leq |f(x_1) - f(c)| + |f(x_2) - f(c)|$$
$$< \frac{1}{2}\epsilon + \frac{1}{2}\epsilon = \epsilon$$

whenever $x_1, x_2 \in (c - \delta, c + \delta)$.

(2) $\implies$ (1) Assume that the given condition is satisfied. Now, letting $x_2 = c$ and replacing x_1 by x in the given condition, it follows that, for every $\epsilon > 0$, there exists $\delta > 0$ such that

$$|f(x) - f(c)| < \epsilon \text{ whenever } x \in (c - \delta, c + \delta),$$

i.e.,

$$|f(x) - f(c)| < \epsilon \text{ whenever } |x - c| < \delta.$$

Thus, f is continuous at c. This completes the proof.

Theorem 5.11.3. *The following are equivalent:*

(1) *A function f is continuous on $\mathbb{R}$;*

(2) *For every open set $O \in \mathbb{R}$, $f^{-1}(O)$ is open in $\mathbb{R}$.*

Proof. (1) $\Longrightarrow$ (2) Let f be continuous on $\mathbb{R}$ and let O be any open set in $\mathbb{R}$. If $f^{-1}(O)$ is empty, then it is open, and so we have the required result.

If $f^{-1}(O)$ is non-empty, then let $c \in f^{-1}(O)$. Then, $f(c) \in O$. Since O is open, there exists $\epsilon > 0$ such that

$$(f(c) - \epsilon, f(c) + \epsilon) \subset O.$$

Since f is continuous on $\mathbb{R}$, it is continuous at $c \in \mathbb{R}$. So, there exists $\delta > 0$ such that

$$|f(x) - f(c)| < \epsilon \text{ whenever } |x - c| < \delta,$$

i.e.,

$$f(x) \in (f(c) - \epsilon, f(c) + \epsilon) \text{ whenever } x \in (c - \delta, c + \delta)$$

i.e.,

$$f(x) \in O \text{ whenever } x \in (c - \delta, c + \delta),$$

i.e.,

$$x \in f^{-1}(O) \text{ whenever } x \in (c - \delta, c + \delta).$$

So, we have

$$(c - \delta, c + \delta) \subset f^{-1}(O).$$

Thus, $f^{-1}(O)$ is necessarily open as c is an arbitrary point.

(2) $\Longrightarrow$ (1) Let $f^{-1}(O)$ is open whenever O is open. Let c be an arbitrary point in $\mathbb{R}$. Let $\epsilon > 0$ be given. Since $(f(c) - \epsilon, f(c) + \epsilon)$ is an open set containing $f(c)$, we have $f^{-1}((f(c) - \epsilon, f(c) + \epsilon))$ must also be an open set containing c. So, we can choose $\delta > 0$ such that

$$(c - \delta, c + \delta) \subset f^{-1}((f(c) - \epsilon, f(c) + \epsilon)),$$

i.e.,

$$f((c - \delta, c + \delta)) \subset (f(c) - \epsilon, f(c) + \epsilon).$$

Hence, we have

$$f(x) \in (f(c) - \epsilon, f(c) + \epsilon) \text{ whenever } x \in (c - \delta, c + \delta).$$

Thus, for every $\epsilon > 0$, there exists a $\delta > 0$ such that

$$|f(x) - f(c)| < \epsilon \text{ whenever } |x - c| < \delta.$$

So, f is continuous at the arbitrary point $c \in \mathbb{R}$. Therefore, f is continuous on $\mathbb{R}$. This completes the proof.

Corollary 5.11.4. *The following are equivalent:*

(1) *A function f is continuous on $\mathbb{R}$;*

(2) *For every closed set $C \in \mathbb{R}$, $f^{-1}(C)$ is closed in $\mathbb{R}$.*

Proof. (1) $\Longrightarrow$ (2) Let f be continuous on $\mathbb{R}$ and let C be any closed set in $\mathbb{R}$. Then, $\mathbb{R} - C$ is open in $\mathbb{R}$, and so, by Theorem 5.11.3, $f^{-1}(\mathbb{R} - C)$ is open in $\mathbb{R}$, i.e., $\mathbb{R} - f^{-1}(C)$ is open in $\mathbb{R}$. Hence, $f^{-1}(C)$ is closed in $\mathbb{R}$.

(2) $\Longrightarrow$ (1) Let $f^{-1}(C)$ be closed in $\mathbb{R}$ whenever C is closed in $\mathbb{R}$. Let O be any open set in $\mathbb{R}$. Then, $\mathbb{R} - O$ is closed in $\mathbb{R}$. So, $f^{-1}(\mathbb{R} - O)$ is closed in $\mathbb{R}$, i.e., $\mathbb{R} - f^{-1}(O)$ is closed in $\mathbb{R}$. Thus, $f^{-1}(O)$ is open in $\mathbb{R}$ whenever O is open in $\mathbb{R}$. Hence, by Theorem 5.11.3, f is continuous on $\mathbb{R}$. This completes the proof.

Example 5.11.1. (1) Every constant function is continuous on $\mathbb{R}$;

(2) The identity function is continuous on $\mathbb{R}$;

(3) The following function:

$$f(x) = a_n x^n + a_{n-1} x^{n-1} + \cdots + a_0$$

is continuous on $\mathbb{R}$;

(4) The functions $f(x) = \sin x$ and $g(x) = \cos x$ are continuous on $\mathbb{R}$;

(5) If $a > 0$, then a^x is continuous for every $x \in \mathbb{R}$;

(6) The logarithmic function $\log_a x$ is continuous on $(0, \infty)$ for a positive non-unity a;

(7) The function x^a is continuous on $(0, \infty)$ for every $a \in \mathbb{R}$;

(8) For every $x, y \in \mathbb{R}$, if a function f satisfies the following:

$$|f(x) - f(y)| \leq \alpha |x - y|,$$

where $\alpha > 0$, then f is continuous on $\mathbb{R}$.

5.12 Properties of Continuous Functions

Theorem 5.12.1. *A continuous function f on a closed bounded interval I is bounded and attains its bounds.*

Proof. Let f be a continuous function on $I = [a, b]$. Suppose that f is not bounded above. Then, for every $n \in \mathbb{N}$, there exists a point $x_n \in [a, b]$ such that $f(x_n) > n$. By the Bolzano–Weierstrass Theorem, the bounded sequence $\{x_n\}$ has a subsequence $\{x_{n_i}\}$ which converges to $c \in [a, b]$. It follows that $\{f(x_{n_i})\}$ is unbounded, but it should converge to $f(c)$ by the continuity of f.

This a contradiction. Thus, f is bounded above. Similarly, it can be proved that f is bounded below.

Now, we show that f attains its bounds. Let K be the supremum of the set $A = \{f(x) : x \in [a,b]\}$. Then, for every $n \in \mathbb{N}$, there exists a point $x_n \in [a,b]$ such that $|K - f(x_n)| < \frac{1}{n}$. (If no such point exists, then $K - \frac{1}{n}$ would be an upper bound of the set A.) The bounded sequence $\{x_n\}$ has a subsequence $\{x_{n_i}\}$ which converges to a point $\alpha \in [a,b]$. Also, we have $|K - f(x_{n_i})| < \frac{1}{n_i}$. Hence, by the continuity of f, it follows that

$$\lim_{n_i \to \infty} f(x_{n_i}) = K = f(\alpha).$$

Thus, there exists a point $\alpha \in [a,b]$ such that $f(\alpha) = K$. In a similar way, it can be proved that f attains its infimum. This completes the proof.

Theorem 5.12.2. *If a function f is continuous on the interval $[a,b]$ and $f(a)$ and $f(b)$ are of opposite signs, then $f(x) = 0$ at some point $x \in (a,b)$.*

Proof. Let us suppose that $f(a) < 0$ and $f(b) > 0$. Let

$$A = \{y \in [a,b] : f(z) \le 0, z \in [a,y]\}.$$

Clearly, A is non-empty and bounded since $a \in A$ and $A \subseteq [a,b]$. So, A has the supremum, say c, by the Completeness Axiom. If possible, suppose that $f(c) > 0$. If we chose $f(c) = \epsilon$, then the continuity of f assures the existence of $\delta > 0$ such that

$$|f(x) - f(c)| < \epsilon \quad \text{whenever} \quad |x - c| < \delta,$$

i.e.,

$$-\epsilon + f(c) < f(x) < \epsilon + f(c) \quad \text{whenever} \quad c - \delta < x < c + \delta,$$

i.e.,

$$f(x) > 0 \quad \text{whenever} \quad x \in (c - \delta, c + \delta).$$

Since c is the supremum of A, there exists $d \in A$, where $c - \delta < d \le c$ such that $f(d) \le 0$. But, from the above relation, $f(d) > 0$, which a contradiction. Hence, $f(c) \ngtr 0$.

Next, if possible, suppose that $f(c) < 0$. If we choose $f(c) = -\epsilon$, then, proceeding as before, we have $f(x) < 0$ whenever $x \in (c - \delta, c + \delta)$ for some $\delta > 0$. So, we have $(c - \delta, c + \delta) \subseteq A$. If we choose $\delta' < \delta$, then $c + \delta' \in A$, which is a contradiction since c is the supremum of A. Hence, $f(c) \nless 0$. It follows that $f(c) = 0$. Since c cannot be equal to either a or b, we have $f(x) = 0$ at some at some point $x \in (a,b)$. This completes the proof.

Corollary 5.12.3. *If f is continuous on $[a,b]$ and α is a real number lying between $f(a)$ and $f(b)$, then f must assume the value α at some point between a and b.*

Proof. Consider the following function:

$$g(x) = f(x) - \alpha$$

and apply Theorem 5.12.3.

Theorem 5.12.4. *If f is a continuous function on $[a, b]$ and $f(x) \in [a, b]$ for every $x \in [a, b]$, then there exists a point $\alpha \in [a, b]$ such that $f(\alpha) = \alpha$ (such a point α is called a fixed point of the function f).*

Proof. We can assume that $f(a) > a$ and $f(b) < b$, as $f(x) \in [a, b]$ for every $x \in [a, b]$, and if $f(a) = a$ or $f(b) = b$, then the theorem is proved. Let us consider the function $g(x) = f(x) - x$, for all $x \in [a, b]$. Then g must be a continuous function on $[a, b]$ being the difference of two continuous functions on $[a, b]$. Since $g(a) > 0$ and $g(b) < 0$, we have $g(\alpha) = 0$ for some $\alpha \in [a, b]$, by Corollary 5.12.3. Thus, $f(\alpha) = \alpha$, for some $\alpha \in [a, b]$, i.e., f has a fixed point on $[a, b]$. This completes the proof.

5.13   Uniform Continuity

The notion of continuity of a function is local in nature in the sense that it is defined for individual points of the domain. Even when we check continuity on the whole domain, we do it by checking continuity at every point of the domain individually. In contrast, the notion of uniform continuity of function is global in nature, in the sense that it is defined for pairs of points of the domain rather than individual points and thus applicable over the whole domain.

Let S be a subset of the set $\mathbb{R}$ of real numbers and $f : S \to \mathbb{R}$ be a function.

Definition 5.13.1. (The Uniform Continuity) The function f is said to be *uniformly continuous* on S if, for every $\epsilon > 0$, there exists $\delta > 0$ such that, for every $x, x_0 \in S$ satisfying $|x - x_0| < \delta$, $|f(x) - f(x_0)| < \epsilon$.

If we compare Definition 5.10.2 with Definition 5.13.1, then we shall see that, in case of the continuity, the number δ in general depends on the selection of ϵ and the individual point x_0. However, in case of the uniform continuity, the number δ depends only on ϵ and not on the selection of the point x_0.

Example 5.13.1. Prove that the function $f(x) = x^2 + 1$ is uniformly continuous on $S = (0, 1)$.

Solution. Choose $\epsilon > 0$ and let $\delta = \frac{\epsilon}{2}$. Choose $x_0 \in S$ and $x \in S$. Therefore, we have $0 < x_0 < 1$ and $0 < x < 1$. Assume $|x - x_0| < \delta$. Then, we have

$$|f(x) - f(x_0)| = |(x^2 + 1) - (x_0^2 + 1)| = |x^2 - x_0^2|$$
$$= |x + x_0||x - x_0|$$
$$\leq (1 + 1)|x - x_0| < 2\delta$$
$$= \epsilon.$$

Hence, the function f is uniformly continuous on $(0, 1)$.

In the solution of Example 5.13.1, the function f satisfies the inequality of the following form:

$$|f(x) - f(y)| \leq K|x - y| \quad \text{for every} \quad x, y \in S. \tag{5.17}$$

The inequality of the form (5.17) is called the *Lipschitz inequality* and the constant K is called the *corresponding Lipschitz constant*.

Theorem 5.13.1. *If a function f satisfies (5.17) for every $x, y \in S$, then f must be uniformly continuous on S.*

Proof. Let x and y be two arbitrary points of S. Let $\epsilon > 0$ be given. Taking $\delta = \frac{\epsilon}{K}$, we get

$$|x - y| < \delta \Rightarrow |f(x) - f(y)| \leq K|x - y| < K\delta = \epsilon.$$

This completes the proof.

Example 5.13.2. Show that the function $f(x) = x^2$ is continuous but not uniformly continuous on $(0, \infty)$.

Solution. First, we show that f is continuous on $(0, \infty)$. Choose $\epsilon > 0$ and $x_0 \in (0, \infty)$ be any point. Let $\delta = \min\{1, \frac{\epsilon}{2x_0+1}\}$. Choose $x \in (0, \infty)$. Assume $|x - x_0| < \delta$. Therefore, $|x - x_0| < 1$ and so $x < x_0 + 1$. Then, we have

$$|f(x) - f(x_0)| = |x^2 - x_0^2| = (x + x_0)|x - x_0| < (2x_0 + 1)|x - x_0|$$
$$< (2x_0 + 1)\delta < (2x_0 + 1)\frac{\epsilon}{(2x_0 + 1)}$$
$$= \epsilon.$$

Thus, f is continuous on $(0, \infty)$.

Next, we show that f is not uniformly continuous on $(0, \infty)$. Let $\epsilon = 1$ and choose $\delta > 0$. Let $x_0 = \frac{3}{\delta}$ and $x = x_0 + \frac{\delta}{6}$. Then, we have

$$|x - x_0| = \frac{\delta}{6} < \delta,$$

but

$$|f(x) - f(x_0)| = |x^2 - x_0^2| = \left|\left(\frac{3}{\delta} + \frac{\delta}{6}\right)^2 - \left(\frac{3}{\delta}\right)^2\right|$$
$$= 1 + \frac{\delta^2}{6^2} > 1 = \epsilon.$$

Hence, f is not uniformly continuous on $(0, \infty)$.

It is clear that the uniform continuity implies the continuity, but the converse is not always true as seen from Example 5.13.2.

If we look at the solutions of Examples 5.13.1 and 5.13.2, respectively, then we must be comfortable in saying that the function $f(x) = x^2$ is uniformly continuous on $[a, b]$ for every $a > 0$. In fact, for every $\epsilon > 0$, we have the condition of the uniform continuity for this function by letting $\delta = \frac{\epsilon}{2b}$. It can be proved in general that every continuous function on any closed bounded interval is uniformly continuous.

Theorem 5.13.2. *If a function f is continuous on a closed bounded interval I, then it is uniformly continuous on I.*

Proof. Suppose to the contrary that f is continuous but not uniformly continuous on I. Then, there exists $\epsilon > 0$ such that, for every $\delta > 0$, there are $x, y \in I$ for which

$$|x - y| < \delta, \quad |f(x) - f(y)| \geq \epsilon.$$

So, there exist $\epsilon_1 > 0$ and $x_n, y_n \in I$ such that

$$|x_n - y_n| < \frac{1}{n}, \quad |f(x_n) - f(y_n)| \geq \epsilon_1 \quad \text{for every } n = 1, 2, \ldots. \quad (5.18)$$

Since $\{x_n\} \subset I$, which is bounded and closed, it follows that $\{x_n\}$ has a subsequence, say $\{x_{n_i}\}$, which converges, as $i \to \infty$, to a point $a \in I$.

Similarly, the sequence $\{y_{n_i}\}$ has a subsequence, say $\{y_{n_{i_j}}\}$, which converges as $j \to \infty$, to a point $b \in I$. Since $\{x_{n_{i_j}}\}$ also converges to $a \in I$ as $j \to \infty$ and f is continuous on I, it follows from (5.18) that $a = b$ and $|f(a) - f(b)| \geq \epsilon_1$, i.e., $f(a) \neq f(b)$, which is a contradiction. Hence, f must be uniformly continuous on I.

Now, the readers should be able to develop a proof for the following result giving some equivalent conditions saying that $f : S \to \mathbb{R}$, $S \subseteq \mathbb{R}$ is not uniformly continuous on S.

Proposition 5.13.3. *Let $f : S \to \mathbb{R}$, $S \subseteq \mathbb{R}$, be a function. Then, the following conditions are equivalent:*

(1) *f is not uniformly continuous on S;*

(2) *There exists $\epsilon_1 > 0$ such that, for every $\delta > 0$, there exist points $x, y \in S$ such that*

$$|x - y| < \delta, \quad |f(x) - f(y)| \geq \epsilon_1;$$

(3) *There exist $\epsilon_2 > 0$ and the sequences $\{x_n\}$, $\{y_n\}$ in S such that*

$$\lim_{n \to \infty} |x_n - y_n| = 0$$

and

$$|f(x_n) - f(y_n)| \geq \epsilon_2 \quad \text{for every } n = 1, 2, \ldots.$$

As an immediate consequence, we have the following result giving a sequential criterion for the uniform continuity:

Proposition 5.13.4. *The following are equivalent:*

(1) *A function $f : S \to \mathbb{R}$, $S \subseteq \mathbb{R}$, is uniformly continuous on S;*

(2) *Whenever sequences $\{x_n\}$ and $\{y_n\}$ in S are such that the sequence $\{x_n - y_n\}$ converges to 0, the sequence $\{f(x_n) - f(y_n)\}$ converges to 0.*

Example 5.13.3. Show that the function $f(x) = \sin \frac{1}{x}$ is not uniformly continuous on $(0, 1]$.

Solution. Let us define two sequences $\{x_n\}$ and $\{y_n\}$ in $(0, 1]$ by

$$x_n = \frac{1}{2n\pi}, \quad y_n = \frac{1}{2n\pi + \frac{\pi}{2}} \quad \text{for every } n = 1, 2, \ldots.$$

Then, $|x_n - y_n| \to 0$ as $n \to \infty$, but

$$|f(x_n) - f(y_n)| = \sin\left(2n\pi + \frac{\pi}{2}\right) - \sin 2n\pi = 1 \quad \text{for every } n = 1, 2, \ldots.$$

Hence, by Proposition 5.13.4, the function $\sin \frac{1}{x}$ is not uniformly continuous on $(0, 1]$. However, it is continuous on $(0, 1]$.

5.14 Continuity and Uniform Continuity in Metric Spaces

Definition 5.14.1. Let (X, d_1) and (Y, d_2) be two metric spaces.

(1) A function $f : X \to Y$ is said to be *continuous* at a point $x_0 \in X$ if, for every $\epsilon > 0$, there exists $\delta > 0$ such that, for every $x \in X$,

$$d_1(x, x_0) < \delta \implies d_2(f(x), f(x_0)) < \epsilon;$$

(2) The mapping f is said to be *continuous* on X if it is continuous at every point of X.

Example 5.14.1. If (X, d) is a metric space, then the identity mapping $I : X \to X$ is continuous on X.

Example 5.14.2. The square root function $f(x) = \sqrt{x}$ is continuous on $[0, \infty)$.

Solution. Let $x, \epsilon > 0$ and $\delta := \epsilon\sqrt{x}$ (for $x = 0$, choose $\delta = \epsilon^2$). Then, we have

$$|x - y| < \delta \Rightarrow |\sqrt{x} - \sqrt{y}| < \frac{\delta}{\sqrt{x} + \sqrt{y}} \leq \frac{\epsilon}{1 + \sqrt{\frac{y}{x}}} < \epsilon.$$

Theorem 5.14.1. *Let (X, d_1) and (Y, d_2) be two metric spaces. Then, the following are equivalent:*

(1) *A mapping $f : X \to Y$ is continuous at a point $x_0 \in X$;*

(2) *For every sequence $\{x_n\}$ in X,*

$$x_n \to x_0 \implies f(x_n) \to f(x_0).$$

Proof. (1) $\implies$ (2) Let f be a continuous mapping at a point $x_0 \in X$. Then, for every $\epsilon > 0$, there exists $\delta > 0$ such that, for every $x \in X$,

$$d_1(x, x_0) < \delta \implies d_2(f(x), f(x_0)) < \epsilon.$$

Let $\{x_n\}$ be a sequence in X such that $x_n \to x_0$ as $n \to \infty$. Then, there exists a positive integer n_0 such that

$$d_1(x_n, x_0) < \delta \quad \text{for every} \ \ n \geq n_0.$$

Hence, for every $n \geq n_0$, we have

$$d_2(f(x_n), f(x_0)) < \epsilon.$$

Therefore, we have

$$x_n \to x_0 \implies f(x_n) \to f(x_0).$$

(2) $\implies$ (1) Assume that, for every sequence $\{x_n\}$ in X,

$$x_n \to x_0 \implies f(x_n) \to f(x_0).$$

Suppose that T is not continuous at $x_0 \in X$. Then, there exists $\epsilon > 0$ such that, for every $\delta > 0$, there exists $x \in X_1$ with $x \neq x_0$ satisfying

$$d_1(x, x_0) < \delta, \quad \text{but} \ \ d_2(f(x), f(x_0)) \geq \epsilon.$$

In particular, for $\delta = \frac{1}{n}$, there exists $x_n \in X$ satisfying

$$d_1(x_n, a_0) < \frac{1}{n}, \quad \text{but} \ \ d_2(f(x_n, f(x_0))) \geq \epsilon.$$

Then, we have clearly $x_n \to x_0$, but $\{f(x_n)\}$ does not converge to $T(x_0)$, which contradicts our assumption $f(x_n) \to f(x_0)$. Thus, T is continuous at $a_0 \in X$. This completes the proof.

Theorem 5.14.2. *Let X, Y and Z be any arbitrary sets. If $f : X \to Y$ and $g : Y \to Z$ are two continuous mappings, then the composition $g \circ f : X \to Z$ of two mappings f and g is continuous on X.*

Proof. Now, we have to show that the composition $g \circ f : X \to Z$ is also a continuous mapping on X. For this, assume that $x_n \to x_0$ in X. Then, by the continuity of S, we have $f(x_n) \to f(x_0)$ in Y. Since g is also continuous on Y, by the continuity of g, we have

$$(g \circ f)(x_n) = g(f(x_n)) \implies g(f(x_0)) = (g \circ f)(x_0)$$

in Z. Thus, $g \circ f : X \to Z$ is also continuous on X. This completes the proof.

Theorem 5.14.3. *The distance function $d : X \times X \to \mathbb{R}$ is also continuous.*

Proof. Suppose that $u_n \to u$ and $v_n \to v$ in X. Then, by using the triangular inequality, we have

$$|d(u_n, v_n) - d(u, v)| \le |d(u_n, v_n) - d(u, v_n)| + |d(u, v_n) - d(u, v_n)|$$
$$\le d(u_n, u) - d(v_n, v) \to 0,$$

which implies that $d(u_n, v_n) \to d(u, v)$ as $n \to \infty$. Hence, d is continuous. This completes the proof.

Definition 5.14.2. Let (X, d_1) and (Y, d_2) be two metric spaces. A mapping $f : X \to Y$ is said to be *uniformly continuous* on X if, for every $\epsilon > 0$, there exists $\delta > 0$ (depending only on ϵ) such that, for every $x, y \in X$,

$$d_1(x, y) < \delta \implies d_2(f(x), f(y)) < \epsilon.$$

Note that every uniform continuous function is continuous, but converse is not true in general. For example, a mapping $f : X \to Y$ is uniformly continuous if X is discrete metric space and Y is any metric space.

Example 5.14.3. (1) Consider a real-valued function $f : [-1, 1] \to \mathbb{R}$ defined by

$$f(x) = x^2 \text{ for every } x \in [-1, 1]$$

and let $x, y \in [-1, 1]$. Then, we have

$$d(f(x), f(y)) = |f(x) - f(y)| = |x^2 - y^2|$$
$$= |x - y| \cdot |x + y| < \epsilon$$

whenever $|x - y| < \frac{1}{2}\epsilon = \delta$, where δ is independent of the choice of x and y. Thus, for every ϵ, there exists $\delta = \frac{1}{2}\epsilon$ such that, for every $x, y \in [-1, 1]$,

$$d(f(x), f(y)) < \epsilon$$

whenever $d(x, x_0) < \delta$;

(2) Now, if we consider the same function $f : \mathbb{R} \to \mathbb{R}$ defined by $f(x) = x^2$ for every $x \in \mathbb{R}$. Then, for every $x, y \in \mathbb{R}$, we have

$$d(f(x), f(y)) = |x^2 - y^2| = |x - y| \cdot |x + y| < \epsilon$$

whenever $|x - y| < \frac{\epsilon}{|x+y|} = \delta$, where δ depends on ϵ and y. So, as in (1), δ is independent of the choice of x and y, and so f is uniformly continuous. But, as in (2), δ depends on ϵ and y, and hence f is only continuous but not uniformly continuous.

Theorem 5.14.4. *Let (X, d_1) and (Y, d_2) be metric spaces and $f : X \to Y$ be a uniformly continuous function. If $\{x_n\}$ is a Cauchy sequence in X, then $\{f(x_n)\}$ is also a Cauchy sequence in Y.*

Proof. Since f is uniformly continuous on X, for every $\epsilon > 0$, there exists $\delta > 0$ (only depends on ϵ) such that, for every $x_1, x_2 \in X$,

$$d_1(x_1, x_2) < \delta \implies d_2(f(x_1), f(x_2)) < \epsilon.$$

In particular, we have

$$d_1(x_n, x_m) < \delta \implies d_2(f(x_n), f(x_m)) < \epsilon.$$

Since $\{x_n\}$ is a Cauchy sequence in X, for every $\delta > 0$, there exists a positive integer n_0 such that

$$d_1(x_n, x_m) < \delta \text{ for every } n, m \geq n_0.$$

So, we obtain

$$d_2(f(x_n), f(x_m)) < \epsilon \text{ for every } n, m \geq n_0.$$

Hence, $\{f(x_n)\}$ is a Cauchy sequence in Y. This completes the proof.

5.15 Exercises

Excercise 5.1 Prove that

(1)If $\lim_{x \to c} f(x) = l > 0$, then $\lim_{x \to c} \log f(x) = \log l$;

(2)Prove that, if $\lim_{x \to c} f(x) = l$, then $\lim_{x \to c} e^{f(x)} = e^l$.

Excercise 5.2 Prove that, if $\lim_{x \to c} f(x) = l \neq 0$, then there exists numbers $k > 0$ and $\delta > 0$ such that

$$|f(x)| > k \text{ whenever } 0 < |x - c| < \delta.$$

Excercise 5.3 Prove that, if a function f is defined on a deleted neighborhood U of a point c and $f(x) \geq 0$ for every $x \in U$, then $\lim_{x \to c} f(x) \geq 0$ provided the limit exists.

Excercise 5.4 Let f and g be defined on a deleted neighborhood U of a point c and let $f(x) \geq g(x)$ for every $x \in U$. Prove that

$$\lim_{x \to c} f(x) \geq \lim_{x \to c} g(x)$$

provided both limits exist.

Hint. Use Exercise 5.3.

Excercise 5.5 Use the Sandwich Theorem to prove that

$$\lim_{x \to 0} x^2 \sin \left(\frac{1}{x} \right) = 0.$$

Hint. See Example 5.2.1.

Excercise 5.6 If $\lim_{x \to c} f(x) = l$, then show that $\lim_{x \to c} |f(x)| = |l|$. Is the converse true?

Hint. Consider the following example for the converse part:

$$f(x) = \begin{cases} 1, & \text{if } x \geq c, \\ -1, & \text{if } x < c. \end{cases}$$

Excercise 5.7 Find the right-hand and left-hand limits of a function defined as follows:

$$g(x) = \begin{cases} \frac{|x-7|}{x-7}, & \text{if } x \neq 7, \\ 0, & \text{if } x = 7. \end{cases}$$

Does $\lim_{x \to 7} g(x)$ exist?

Excercise 5.8 Show that $\lim_{x \to 0} \frac{|x|}{2x}$ does not exist.

Excercise 5.9 Prove that

$$\lim_{x \to \infty} \frac{1}{x + 3} = 0.$$

Excercise 5.10 Evaluate the following:

(1) $\lim_{x \to 0} [x]$, where $[x]$ denotes the greatest integer not greater than x;

(2) $\lim_{x \to 0} \log |x|$;

(3) $\lim_{x \to 0} \frac{1}{\sqrt{|x|}}$;

(4) $\lim_{x \to 0+} \frac{1}{1 + e^{-\frac{1}{x}}}$.

Excercise 5.11 Examine the existence of the following limit in the sense of Definition 5.6.1:

$$\lim_{x \to 0} \frac{1}{x} \sin \frac{1}{x}.$$

Excercise 5.12 Prove that a function f tends to a finite limit as x tends to ∞ if and only if, corresponding to every $\epsilon > 0$, there exists $M > 0$ such that

$$|f(x) - f(y)| < \epsilon \text{ for every } x, y > M.$$

Hint. See Theorem 5.7.1.

Excercise 5.13 Prove that the function $f(x) = \cos x$ is strictly decreasing on the interval $[0, \pi]$.

Hint. Use the cosine difference identity:

$$\cos x_2 - \cos x_1 = -2 \sin \frac{x_2 + x_1}{2} \sin \frac{x_2 - x_1}{2}.$$

Excercise 5.14 Prove that the function f defined as $f(x) = x^{\frac{1}{k}}$ is continuous at every $x > 0$, where k is any natural number.

Hint. Use the following:

$$x - y = \left(x^{\frac{1}{k}} - y^{\frac{1}{k}} \right) \left(x^{\frac{(k-1)}{k}} + x^{\frac{(k-2)}{k}} y^{\frac{1}{k}} + \cdots + y^{\frac{(k-1)}{k}} \right).$$

If $y > \frac{x}{2}$, then we have

$$x^{\frac{(k-1)}{k}} + x^{\frac{(k-2)}{k}} y^{\frac{1}{k}} + \cdots + y^{\frac{(k-1)}{k}} > \frac{k}{2} x^{\frac{(k-1)}{k}}.$$

Let $\epsilon > 0$ and $\delta = \min \left\{ \frac{x}{2}, \frac{k}{2} x^{\frac{(k-1)}{k}} \epsilon \right\}$. Then, we have

$$|x - y| < \delta \implies |x^{\frac{1}{k}} - y^{\frac{1}{k}}| < \delta \cdot \frac{2}{k} x^{\frac{k}{(k-1)}} < \epsilon.$$

Excercise 5.15 Prove that the following function:

$$f(x) = \begin{cases} x \cos \frac{1}{x}, & \text{if } x \neq 0, \\ 0, & \text{if } x = 0, \end{cases}$$

is continuous at $x = 0$.

Excercise 5.16 Discuss the continuity of the following function:

$$f(x) = x - [x]$$

at $x = 2$.

Excercise 5.17 Prove that Dirichlet's function f defined on $\mathbb{R}$ by

$$f(x) = \begin{cases} 1, & \text{if } x \notin \mathbb{Q}, \\ -1, & \text{if } x \in \mathbb{Q}. \end{cases}$$

is discontinuous at every point.

Hint. See Example 5.10.4.

Excercise 5.18 Construct different functions having the removable discontinuity, the discontinuity of the first kind, the discontinuity of the second kind, the mixed discontinuity, and the infinite discontinuity.

Excercise 5.19 Prove that, if f is continuous at c, then $|f|$ is also continuous at c. Is the converse true?

Hint. Consider the following inequality:

$$||f(x)| - |f(c)|| \le |f(x) - f(c)|.$$

See Exercise 5.6 for the converse part.

Excercise 5.20 Prove that, if f and g are continuous functions at c, then the functions $\max\{f, g\}$ and $\min\{f, g\}$ are both continuous at c.

Hint. Use

$$\max\{f, g\} = \frac{1}{2}(f + g) + \frac{1}{2}|f - g|$$

and

$$\min\{f, g\} = \frac{1}{2}(f + g) - \frac{1}{2}|f - g|.$$

Excercise 5.21 Prove that, if a function g is continuous at c and f is continuous at $g(c)$, then the composite function $f \circ g$ is continuous at c.

Excercise 5.22 Prove that functions mentioned in Example 5.11.1 are continuous.

Excercise 5.23 Construct some examples to show that a closed bounded interval is a must for Theorem 5.12.1 to work.

Hint. One may look at the following real valued functions:

(a) $f_1(x) = x^2$ is continuous but not bounded on $[0, \infty)$;

(b) $f_2(x) = \frac{x}{1+x}$ is continuous and bounded on $[0, \infty)$ but does not attain its supremum;

(c) $f_3(x) = x$ is continuous and bounded on $(0, 1)$ but does not attain its supremum and infimum;

(d) $f_4(x) = \frac{1}{x^3}$ is continuous but not bounded on $(0, 1]$.

Excercise 5.24 Prove that the image $f(I)$ of an interval I under a continuous mapping f is also an interval.

Hint. If y lies between $f(a)$ and $f(b)$, then, by the Corollary 5.12.3, we have $f(x) = y$ for some x between a and b.

Excercise 5.25 Prove that the image of a closed and bounded interval $I = [a, b]$ under a continuous mapping f is also closed and bounded.

Hint. Use Theorem 5.12.1.

Excercise 5.26 Prove that, if f is continuous and one-to-one on an interval I, then f is strictly monotonic on I.

Excercise 5.27 Prove that if f is continuous and one-to-one on an interval I, then f^{-1} exists and is continuous and one-to-one on $f(I)$.

Excercise 5.28 Prove that the function $f(x) = \sin x$ is uniformly continuous on $[0, \infty)$.

Hint. Consider the following:

$$|f(x) - f(y)| = |\sin x - \sin y| = \left| 2\sin \frac{x-y}{2} \cos \frac{x+y}{2} \right|$$

and

$$|\sin x| \leq |x|, \quad |\cos x| \leq 1$$

for all $x \in [0, \infty)$.

Excercise 5.29 Prove that the function $f(x) = \frac{1}{x}$ is not uniformly continuous on $]0, 1]$.

Hint. Consider the sequences $\{x_n\}$ and $\{y_n\}$ defined by

$$x_n = \frac{1}{n} \quad \text{and} \quad y_n = \frac{1}{n+1} \quad \text{for every } n = 1, 2, \ldots,$$

respectively, and apply Proposition 5.13.3. Readers may try other methods as well.

Excercise 5.30 Prove that $f(x) = \sin x^2$ is not uniformly continuous on $[0, \infty)$.

Excercise 5.31 Let (X, d) be a metric space and suppose that, for every $i = 1, 2, \ldots, n$, the function $f_i : X \to \mathbb{R}$ is continuous. Prove that the function $f : X \to \mathbb{R}^n$ defined by

$$f(x) = (f_1(x), f_2(x), \ldots, f_n(x)) \quad \text{for every } x \in X$$

is continuous.

Hint. Let $x_0 \in X$ be an arbitrary point. First, use the continuity of each f_i so that, corresponding to every $\epsilon > 0$, there exists $\delta_i > 0$ for every $i = 1, 2, \ldots, n$ such that, for every $x \in X$ and $i = 1, 2, \ldots, n$,

$$d(x, x_0) < \delta_i \implies |f_i(x) - f_i(x_0)| < \frac{\epsilon}{\sqrt{n}}.$$

Then, $d(x, x_0) < \delta = \min\{\delta_1, \delta_2, \cdots, \delta_n\}$ implies that

$$d(f(x), f(x_0)) = \left[(f_1(x) - f_1(x_0))^2 + \cdots + (f_n(x) - f_n(x_0))^2\right]^{\frac{1}{2}} < \epsilon.$$

Excercise 5.32 Give an example of a function defined on a metric space which is not uniformly continuous.

Hint. Let $f : X \to Y$ be a function defined by

$$f(x) = \frac{1}{x} \quad \text{for every} \ \ x \in X,$$

where $X = (0, 1)$ is a metric space with the Euclidean metric $d(x, y) = |x - y|$ for every $x, y \in (0, 1)$ and $Y = \mathbb{R}$ is a metric spaces with the Euclidean metric $d(x, y) = |x - y|$ for every $x, y \in \mathbb{R}$. Then, f is not uniformly continuous on X.

Let $\epsilon = \frac{1}{2}$. Choose δ be any positive number. Let $x = \frac{1}{n}$ and $y = \frac{1}{n+1}$, where n is a positive integer such that $n > \frac{1}{\delta}$. Then, we have

$$|x - y| = \left|\frac{1}{n} - \frac{1}{n+1}\right| = \frac{1}{n(n+1)} < \frac{1}{n} < \delta,$$

but

$$|f(x) - f(y)| = |n - (n+1)| = 1 > \epsilon.$$

Readers should also try other examples.

Excercise 5.33 Let (X, d) be a metric space and A be a subset of X. Prove that the function $f : X \to \mathbb{R}$ defined by

$$f(x) = d(x, A) \quad \text{for every} \ \ x \in X$$

is uniformly continuous.

Hint. For every $a \in A$ and $x \in X$, apply the triangular inequality to obtain

$$\inf_{a \in A} d(x, a) \leq d(x, y) + \inf_{a \in A} d(y, a).$$

Thus, we have

$$d(x, A) \leq d(x, y) + d(y, A)$$

and so

$$d(x, A) - d(y, A) \leq d(x, y) \quad \text{for every} \ \ x, y \in X.$$

Interchange x and y to obtain

$$d(y, A) - d(x, A) \leq d(y, x) = d(x, y).$$

Hence, we have

$$|d(x, A) - d(y, A)| \leq d(x, y).$$

Finally, consider the following inequality:

$$|f(x) - f(y)| = |d(x, A) - d(y, A)| \leq d(x, y).$$

6

Connectedness and Compactness

This chapter deals with the notion of connectedness in metric space, Intermediate Value Theorem, components, compactness in metric space, Finite Intersection Property, totally bounded sets, the Bolzano–Weierstrass Theorem, sequentially compact spaces, and the Heine–Borel Theorem. Solved examples and chapter-end exercises are also incorporated.

6.1   Connectedness

First, we introduce the definition of a separated set.

Definition 6.1.1. Let (X, d) be a metric space. Two non-empty subsets A and B of X are said to be *separated* if

$$A \bigcap \overline{B} = \emptyset, \quad \overline{A} \bigcap B = \emptyset.$$

Example 6.1.1. (1) In the Euclidean metric space $\mathbb{R}$, the sets $A = (1, 2)$ and $B = (2, 3)$ are separated, but the sets $A = (1, 2]$ and $B = (2, 3)$ are not separated;

(2) In the discrete metric space $\mathbb{R}$, the sets $A = (1, 2]$ and $B = (2, 3)$ are separated.

Solution. (1) We have

$$A \bigcap \overline{B} = (1, 2) \bigcap [2, 3] = \emptyset$$

and

$$\overline{A} \bigcap B = [1, 2] \bigcap (2, 3) = \emptyset,$$

and so the sets $A = (1, 2)$ and $B = (2, 3)$ are separated, but

$$A \bigcap \overline{B} = (1, 2] \bigcap [2, 3] \neq \emptyset$$

and

$$\overline{A} \bigcap B = [1, 2] \bigcap (2, 3) = \emptyset,$$

and so the sets $A = (1, 2]$ and $B = (2, 3)$ are not separated.

(2) In the discrete metric space $\mathbb{R}$, we have $A = \overline{A}$ and $B = \overline{B}$, and so the sets $A = (1, 2]$ and $B = (2, 3)$ are separated.

Note that, if the sets A and B are separated in a metric space (X, d), then they are disjoint since we have

$$A \bigcap B \subseteq A \bigcap \overline{B} = \emptyset,$$

but two disjoint sets A and B need not be separated. For example, the sets $A = (1, 2]$ and $B = (2, 3)$ are disjoint but not separated in the usual metric space $\mathbb{R}$.

Let (X, d) be a metric space and, for all subsets A, B of X, define

$$d(A, B) = \inf\{d(x, y) : x \in A, \; y \in B\}.$$

Then, if $d(A, B) > 0$, then A and B are separated. In fact, let $d(A, B) = \lambda$. Then, we have

$$d(x, y) \geq \lambda$$

for every $x \in A$ and $y \in B$. If $x \in A$, then the point x cannot be a limit point of B because the open ball $B_{\frac{\lambda}{2}}(x)$ does not contains any points of B. Thus, we have

$$A \bigcap \overline{B} = \emptyset.$$

Similarly, we have

$$\overline{A} \bigcap B = \emptyset.$$

Therefore, A and B are separated.

But, if A and B are separated, then $d(A, B) > 0$ is not true. For example, in the Euclidean metric space $\mathbb{R}$, the sets

$$A = \{x \in \mathbb{R} : x > 0\}, \quad B = \{x \in \mathbb{R} : x < 0\}$$

are separated, but $d(A, B) = 0$.

Theorem 6.1.1. *Let (X, d) be a metric space and A, B be two non-empty subsets of X. Then, we have the following:*

(1) *If A and B are closed, then A and B are separated if and only if A and B are disjoint;*

(2) *If A and B are open, then A and B are separated if and only if A and B are disjoint.*

Proof. (1) We know that two separated sets A and B are disjoint. Let A and B be disjoint. Since A and B are closed, we have

$$A \bigcap \overline{B} = \overline{A} \bigcap B = A \bigcap B = \emptyset.$$

Thus, the sets A and B are separated.

(2) Let A and B be open and disjoint. Then, $X - A$ and $X - B$ are closed, and so

$$X - A = \overline{X - A}, \quad X - B = \overline{X - B}$$

and

$$A \subseteq X - B \implies \overline{A} \subseteq \overline{X - B} = X - B.$$

Similarly, we can show that $\overline{B} \subseteq X - A$. Therefore, we have

$$\overline{A} \subseteq (X - B) \bigcap B = \emptyset, \quad \overline{B} \subseteq (X - A) \bigcap A = \emptyset.$$

Therefore, the sets A and B are separated. This completes the proof.

Theorem 6.1.2. *Let (X, d) be a metric space, A and B be two separated subsets of X, and let $G = A \bigcup B$. Then, we have the following:*

(1) *A and B are open if G is open;*

(2) *A and B are closed if G is closed.*

Proof. (1) Since $\overline{B}$ is closed, $(\overline{B})^c$ is open. Since $G = A \bigcup B$ is open, $(A \bigcup B) \bigcup (\overline{B})^c$ is also open. Since A and B are separated, $A \bigcap \overline{B} = \emptyset$, and so $A \subseteq (\overline{B})^c$. Also, since $B \subseteq \overline{B}$, we have $(\overline{B})^c \subseteq \overline{B}$ and so

$$B \bigcap (\overline{B})^c \subseteq B \bigcap \overline{B} = \emptyset.$$

Therefore, we have

$$(A \bigcup B) \bigcap (\overline{B})^c = [A \bigcap (\overline{B})^c] \bigcup [B \bigcap (\overline{B})^c] = A \bigcup \emptyset = A.$$

Thus, A is open. Similarly, we can prove that B is open.

(2) Since $G = A \cup B$ is closed, we have

$$A \bigcup B = \overline{A \bigcup B} = \overline{A} \bigcup \overline{B}.$$

Since A and B are separated, we have $\overline{A} \bigcap B = \emptyset$. Thus, we have

$$\overline{A} = \overline{A} \cap (\overline{A} \bigcup \overline{B})$$
$$= \overline{A} \bigcap (A \bigcup B)$$
$$= (\overline{A} \bigcap A) \bigcup (\overline{A} \bigcap B)$$
$$= A \bigcup \emptyset = A.$$

Thus, A is closed. Similarly, we can prove that B is closed. This completes the proof.

Definition 6.1.2. (1) A metric space (X, d) is said to be *connected* if it cannot be expressed as the union of two separated sets;

(2) X is said to be *disconnected* if it is not connected. In other words, X is said to be *disconnected* if

$$X = A \bigcup B,$$

where $A \cap \overline{B} = \emptyset$ and $\overline{A} \cap B = \emptyset$;

(3) The sets A and B are said to form a *separation* of X.

From Theorem 6.1.2, note the following:

(1) A metric space (X, d) is connected if it cannot be expressed as the union of two non-empty disjoint closed sets;

(2) A metric space (X, d) is connected if it cannot be expressed as the union of two non-empty disjoint open sets.

Definition 6.1.3. Let (X, d) be a space and Y be a subset of X. Then, Y is said to be *connected* if the subspace (Y, d_Y) with the metric d_Y induced by d is connected. In other words, (Y, d_Y) is said to be *connected* if

$$Y \neq A \bigcup B,$$

where A and B are non-empty disjoint open subsets of Y, that is, $A = G \cap Y$ and $B = H \cap Y$, where G and H are open subsets of X.

Example 6.1.2. The empty set and a singleton set are always connected.

Example 6.1.3. Any subset of $\mathbb{Z}$ (or a discrete metric space (X, d)) is disconnected.

Example 6.1.4. (1) Let $\mathbb{Q}$ be the set of rational numbers with the Euclidean metric $d(x, y) = |x - y|$ for every $x, y \in \mathbb{Q}$. Show that $\mathbb{Q}$ is disconnected. In fact, the sets

$$A = \{x \in \mathbb{Q} : x > \sqrt{2}\}, \quad \mathbb{B} = \{x \in \mathbb{Q} : x < \sqrt{2}\}$$

are open in $\mathbb{Q}$ and non-empty disjoint subsets with

$$A \bigcap \overline{B} = \emptyset, \quad \overline{A} \bigcap B = \emptyset, \quad \mathbb{Q} = A \bigcup \mathbb{B}.$$

Hence, $\mathbb{Q}$ is disconnected;

(2) Similarly, the set $\mathbb{Q}^c$ of irrational numbers also is disconnected.

Note that a subset X of $\mathbb{R}$ is an interval if and only if, for every $x, y \in X$ with $x < y$, we have $(x, y) \subset X$, that is, if, for every $x, y \in X$, $x < z < y$, then $z \in X$.

Theorem 6.1.3. *Let X be a subset X of the Euclidean metric space $\mathbb{R}$. Then, the following are equivalent:*

(1) X *is connected;*

(2) X *is an interval.*

Proof. (1) $\implies$ (2) Suppose that $X \subseteq \mathbb{R}$ is connected. Suppose that X is not an interval. Then, there exists $a, b, c \in \mathbb{R}$ such that $a < c < b$ and $a, b \in X$ but $c \notin X$. So, we can write X as follows:

$$X = \left[X \bigcap (-\infty, c) \right] \bigcup \left[X \bigcap (c, +\infty) \right]$$

and

$$\left[X \bigcap (-\infty, c) \right] \bigcap \left[X \bigcap (c, +\infty) \right] = \emptyset.$$

Since $a \in X \bigcap (-\infty, c)$ and $b \in X \bigcap (c, +\infty)$, it follows that $X \bigcap (-\infty, c)$ and $X \bigcap (c, +\infty)$ are non-empty disjoint open sets in X. Thus, X is the union of two non-empty disjoint open sets, and hence X is disconnected, which contradicts our assumption. Thus, X is an interval.

(2) $\implies$ (1) Suppose that X is an interval, but it is not connected. Then, there exist two non-empty closed sets U and V in X such that

$$X = U \bigcup V, \quad U \bigcap V = \emptyset.$$

Since U and V are non-empty, we can choose $a \in U$ and $b \in V$. Since U and V are disjoint, we have $a \neq b$. We may assume that $a < b$. Since X is an interval, $[a, b] \subseteq X$, and every point in $[a, b]$ is contained in either U or V. Now, we define c as

$$c = \sup([a, b] \bigcap U) = \sup\{u \in U : a \leq u \leq b\}.$$

Then, we obtain $c \in [a, b]$ and so $c \in X$. Since U is closed, by the definition of c, we have $c \in U$, which shows that $c < b$. Again, by the definition of c, $c + \varepsilon \in H$ for every $\varepsilon > 0$ with $c + \varepsilon \leq b$. Since V is closed in X, we have $c \in V$. Thus, c belongs to both U and V, which contradicts our assumption that the sets U and V are disjoint. Hence, Y is connected. This completes the proof.

Corollary 6.1.4. *The real line $\mathbb{R}$ is connected.*

Proof. Since $\mathbb{R} = (-\infty, +\infty)$ is an interval, the conclusion follows by Theorem 6.1.3.

Theorem 6.1.5. *Let (X, d) be a metric space. Then, the following are equivalent:*

(1) X *is connected;*

(2) *The only non-empty subset of X which is both open and closed in X itself.*

Proof. (1) $\Longrightarrow$ (2) Let X be connected. Assume that A is a non-empty proper subset of X which is both open and closed. Let $B = X - A$. Then, clearly $X = A \bigcup B$, and A, B are non-empty disjoint open subsets of X. Hence, X is disconnected, which contradicts our assumption. Thus, X is the only non-empty set which is both open and closed.

(2) $\Longrightarrow$ (1) Let X be the only non-empty set which is both open and closed. Suppose that X is disconnected. Then, $X = A \bigcup B$, where A and B are non-empty disjoint open subsets of X. This implies that $B = X - A$. Since A is open, B is closed. Thus, B is open as well as a closed proper subset of X, which contradicts our assumption. Hence, X is connected. This completes the proof.

Theorem 6.1.6. *If A is a connected subset of a metric space X, then each subset B such that $A \subseteq B \subseteq \overline{A}$ is connected. In particular, $\overline{A}$ is connected.*

Proof. Suppose that B is not connected. Then, there exist two non-empty open sets G and H in B such that

$$B = G \bigcup H, \quad G \bigcap H = \emptyset.$$

Since $A \subseteq B$ and G, H are open in B, we have

$$X = G \bigcap A, \quad Y = H \bigcap A$$

as non-empty open sets in A. Moreover, we have

$$A = B \bigcap A = (G \bigcup H) \bigcap A = (G \bigcap A) \bigcup (H \bigcap A) = X \bigcup Y$$

and

$$X \bigcap Y = (G \bigcap A) \bigcap (H \bigcap A) = (G \bigcap H) \bigcap A = \emptyset \bigcap A = \emptyset,$$

which contradicts our supposition that A is connected. Hence, B is connected.

Especially, since $A \subseteq B \subseteq \overline{A}$, we have $A \subseteq \overline{A} \subseteq \overline{A}$, and so, if we put $B = \overline{A}$, then $\overline{A}$ is connected. This completes the proof.

Theorem 6.1.7. *If $\{U_\mu\}_{\mu \in \Lambda}$ is a family of connected sets in a metric space X such that $\bigcap_{\mu \in \Lambda} U_\mu \neq \emptyset$, then $U = \bigcup_{\mu \in \Lambda} U_\mu$ is connected.*

Proof. Suppose that U is not connected. Then, there exist non-empty sets $A \subseteq U$ and $B \subseteq U$ open in U such that

$$U = A \bigcup B, \quad A \bigcap B = \emptyset.$$

Let $x \in \bigcup_{\mu \in \Lambda} U_\mu \subseteq U = A \bigcup B$. Then, either $x \in A$ or $x \in B$.

Now, assume that $x \in A$. Then, $A \bigcap U_\mu \neq \emptyset$ for each $\mu \in \Lambda$. Moreover, there exists an index μ_0 such that $A \bigcap U_{\mu_0} \neq \emptyset$.

Similarly, we have $B \bigcap U_{\mu_0} \neq \emptyset$. Therefore, the sets $A_0 = A \bigcap U_{\mu_0}$ and $B_0 = B \bigcap U_{\mu_0}$ are non-empty open sets in U_{μ_0} such that

$$
\begin{aligned}
U_{\mu_0} = U \bigcap U_{\mu_0} &= (A \bigcup B) \bigcap U_{\mu_0} \\
&= (A \bigcap U_{\mu_0}) \bigcup (B \bigcap U_{\mu_0}) \\
&= A_0 \bigcup B_0
\end{aligned}
$$

and

$$
\begin{aligned}
A_0 \bigcap B_0 &= (A \bigcap U_{\mu_0}) \bigcap (B \bigcap U_{\mu_0}) \\
&= (A \bigcap B) \bigcap U_{\mu_0} \\
&= \emptyset \bigcap U_{\mu_0} \\
&= \emptyset,
\end{aligned}
$$

which contradicts the hypothesis that U_{μ_0} is connected. Therefore, $U = \bigcup_{\mu \in \Lambda} U_\mu$ is connected. This completes the proof.

Note the following:

(1) Let A, B be non-empty connected subsets of a metric space (X, d). Then, $A \bigcup B$ are connected if and only if there exists $x \in A \bigcup B$ such that $\{x\} \bigcup A$ and $\{x\} \bigcup B$ are connected;

(2) If C is a connected subset of X, then $\overline{C}$ is also connected.

6.2   The Intermediate Value Theorem

In this section, we prove the Intermediate Value Theorem by using Theorem 6.1.3; that is, a subset of the usual metric space $\mathbb{R}$ is connected if and only if it is an interval.

Definition 6.2.1. Let $Y = \{0, 1\}$ and d^* denote the discrete metric on X. The metric space (Y, d^*) is called the *discrete two element space*.

Theorem 6.2.1. *Let (X, d) be a metric space. Then, the following are equivalent:*

(1) *X is disconnected;*

(2) *There exists a continuous mapping from (X, d) onto the discrete two element space (Y, d^*).*

Proof. $(1) \implies (2)$ Let (X, d) be disconnected. Then, there exist disjoint open subsets A and B of X such that

$$
X = A \bigcup B, \quad A \bigcap B = \emptyset.
$$

Define a mapping $T : X \to Y$ by

$$f(x) = \begin{cases} 0, & \text{if } x \in A, \\ 1, & \text{if } x \in B. \end{cases}$$

Obviously, $T^{-1}(\{0\}) = A$ and $T^{-1}(\{1\}) = B$. Therefore, T is onto.

Now, we show that $T : (X, d) \to (Y, d^*)$ is continuous on X. The open subsets of the discrete two element space (Y, d^*) are $\emptyset$, $\{0\}$, $\{1\}$ and $\{0, 1\}$. Clearly, $T^{-1}(\emptyset) = \emptyset$ and $T^{-1}(\{0, 1\}) = X$ are open in (X, d). Moreover, $T^{-1}(\{0\}) = A$ and $T^{-1}(\{1\}) = B$ are open sets in (X, d). Hence, T is continuous on X.

$(2) \implies (1)$ Let $T : (X, d) \to (Y, d^*)$ be continuous on X and onto. Let $A = T^{-1}(\{0\})$ and $B = T^{-1}(\{1\})$. Then, A and B are non-empty disjoint open sets in X such that $X = A \cup B$. Therefore, X is disconnected. This completes the proof.

Theorem 6.2.2. *If (X, d) is a connected metric space and $f : X \to \mathbb{R}$ is a continuous function on X, then $f(X)$ is connected.*

Proof. Note that $f : X \to f(X)$ is a continuous function on X. Suppose that $f(X)$ is not connected. Then, by Theorem 6.2.1, there exists a continuous function $g : f(X) \to Y$, where (Y, d^*) is the discrete two element space. Thus, the composition $g \circ f : X \to Y$ of two continuous functions is also continuous on X. Again, by using Theorem 6.2.1, X is disconnected, that is, not connected, which is a contradiction. Therefore, $f(X)$ is connected. This completes the proof.

By using Theorem 6.2.2, we can prove the Intermediate Value Theorem as follows:

Theorem 6.2.3. (The Intermediate Value Theorem) *If $f : [a, b] \to \mathbb{R}$ is a continuous function on $[a, b]$, then, for every z with $f(x) \leq z \leq f(y)$ or $f(y) \leq z \leq f(x)$, there exists $c \in [a, b]$ such that $f(c) = z$.*

Proof. Since $f : [a, b] \to \mathbb{R}$ is continuous on $[a, b]$, by Theorem 6.2.2, $f([a, b])$ is connected in $\mathbb{R}$ and $f([a, b])$ is an interval. Thus, by the Interval Property, for every $f(x), f(y) \in f([a, b])$ with $f(x) < z < f(y)$, we have $z \in f([a, b])$. Therefore, there exists $c \in [a, b]$ such that $f(c) = z$. This completes the proof.

As an application of the Intermediate Value Theorem, we have the following.

Theorem 6.2.4. *If $I = [-n, n]$ for every $n \geq 1$ and $f : I \to I$ is a continuous function on I, that is, onto, then there exists $c \in (-n, n)$ such that $f(c) = c$.*

Proof. If $f(-n) = -n$ or $f(n) = n$, then we obtain the desired result. So, we assume that $f(-n) > -n$ and $f(n) < n$. Define a function $F : I \to I$ by

$$F(x) = T(x) - x \quad \text{for every } x \in I.$$

Since f is continuous on I, F is also continuous on I. Also, F satisfies $F(-n) = f(-n) + n > 0$ and $F(n) = f(n) - n < 0$. Thus, by the Intermediate Value Theorem, there exists $c \in (-n, n)$ such that $F(c) = 0$, that is, $f(c) = c$. This completes the proof.

6.3 Components

By a *maximal connected subset* C of a metric space (X, d) is meant a subset C such that any set A with $C \subset A$ $(A \neq C)$ is disconnected, that is, not connected.

Definition 6.3.1. A maximal connected subset C of a metric space (X, d) is called the *component* of X. In other words, C is called a *component* of a metric space X if C is connected and it is not properly contained in any larger connected subset of X.

Let (X, d) be a metric space and $x \in X$. The union $C(x)$ of all connected subsets containing x is clearly a maximal connected sunset of X containing x. Such union $C(x)$ is called a *connected component* of x in X.

Example 6.3.1. Let $Y = [-1, 0) \cup (0, 1] \subset \mathbb{R}$. Find the components of Y.

Solution. Clearly, the components of Y are the two sets $[-1, 0)$ and $(0, 1]$.

Theorem 6.3.1. *If C is a component C of a metric space (X, d), then C is closed.*

Proof. Suppose that the component C is not closed. The C is properly contained in $\overline{C}$. By Theorem 6.1.5, $\overline{C}$ is connected. Thus, C is properly contained in the connected subset $\overline{C}$ of X, which contradicts the maximality of C. This completes the proof.

Theorem 6.3.2. *If C is a connected subset of a metric space (X, d) which is both open and closed, then C is a component of X.*

Proof. Suppose that A is a component of X. Then, every connected subset of X is contained in A; that is, since C is connected in X, $C \subseteq A$.

Now, we show $A = C$. Let $A \neq C$, then $A \cap C$ is a non-empty open set in A since C is open in X. Since C is closed in X, it follows that $X - C$ is open and so $A \cap (X - C)$ is open in A. Also, we have

$$(A \cap C) \cap [A \cap (X - C)] = A \cap [C \cap (X - C)] = A \cap \emptyset = \emptyset$$

and

$$(A \cap C) \cup [A \cap (X - C)] = A \cap [C \cup (X - C)] = A \cap X = A,$$

which implies that A is the union of two disjoint open subsets $A \cap C$ and $C \cap (X - C)$. Therefore, A is not connected, which is a contradiction. So, we have $C = A$. This completes the proof.

Theorem 6.3.3. *Each point of a metric space (X, d) is contained in exactly one component of X.*

Proof. Let $x \in X$ be arbitrary, and consider a family $\{C_\alpha\}_{\alpha \in \Lambda}$ of all connected subsets of X with $x \in C_\alpha$ for all $\alpha \in \Lambda$, where Λ is the index set. Since the singleton $\{x\}$ is connected, the family $\{C_\alpha\}_{\alpha \in \Lambda}$ is non-empty and $\bigcap_{\alpha \in \Lambda} C_\alpha \neq \emptyset$. By Theorem 6.1.6, $C = \bigcup_{\alpha \in \Lambda} C_\alpha$ is connected in X which contains x.

Suppose that C^* is another component of X containing x. Since C is the union of all connected subsets of X, $C^* \subseteq C$. Since C^* is a maximal connected subset of X, we have $C^* = C$. This completes the proof.

6.4   Compactness

Before dealing with compact sets, we need to know some properties of bounded sets.

Definition 6.4.1. A non-empty subset A of a metric space (X, d) is said to be *bounded* if there exists a number $r > 0$ such that

$$d(x, y) \leq r \quad \text{for every} \quad x, y \in A.$$

The least upper bound

$$\delta(A) = \sup\{d(x, y) : x, y \in A\}$$

is called the *diameter* of the set A. Usually, we say that the set A is *bounded* if

$$\delta(A) \leq r < \infty.$$

Otherwise, we say that the set A is *unbounded*.

Theorem 6.4.1. *Any subset C of a bounded set A in a metric space (X, d) is bounded.*

Proof. Suppose that A be a bounded set with $d(x, y) \leq r$ for every $x, y \in A$. In particular, since $c \subseteq A$, also we have $d(x, y) \leq r$ for every $x, y \in C$. Thus, C is also bounded. This completes the proof.

Theorem 6.4.2. *The union of a finite number of bounded sets is bounded.*

Proof. Let $A_1, A_2, \ldots, A_N$ be bounded sets with diameters $\delta_1, \delta_2, \ldots, \delta_N$, respectively. Let $\delta := \max\{\delta_n : n = 1, 2, \ldots, N\}$. Pick each point $a_n \in A_n$ for every $n = 1, 2, \ldots, N$, and take

$$\delta^* := \max\{d(a_n, a_m) : n, m = 1, 2, \ldots, N\}.$$

Now, for every $x, y \in \bigcup_{n=1}^{N} A_n$, that is, $x \in A_i$, $y \in A_j$, by using the triangle inequality, we have

$$
\begin{aligned}
d(x, y) &\le d(x, a_i) + d(a_i, a_j) + d(a_j, y) \\
&\le \delta_i + \delta^* + \delta_j \\
&\le 2\delta + \delta^*,
\end{aligned}
$$

that is, $\{d(x, y) : x, y \in \bigcup_{n=1}^{N} A_n\}$ is upper bounded and so $\bigcup_{n=1}^{N} A_n$ is bounded. This completes the proof.

Theorem 6.4.3. *Let (X, d) be a metric space. Then, the following are equivalent:*

(1) *B is a bounded set in a metric space (X, d);*

(2) *B is a subset of a ball $B_r(a) = \{x \in X : d(a, x) < r\}$.*

Proof. (1) $\Longrightarrow$ (2) Clearly, balls (and their subsets) are bounded, that is, we have

$$d(x, y) \le d(x, a) + d(a, y) < r + r = 2r \quad \text{for every} \quad x, y \in B_r(a).$$

(2) $\Longrightarrow$ (1) If a non-empty set is bounded by a constant R, pick a point $a \in X$ and $b \in B$ to have $x \in B_r(a)$; that is, for every $x \in B$, we have

$$d(x, a) \le d(x, b) + d(b, a) < R + 1 + d(b, a) =: r,$$

and so B is a subset of a ball $B_r(a)$. This completes the proof.

Note that if B is a bounded set in a metric space (x, d) and $f : B \to X$ is a continuous mapping, then $f(B)$ need not be bounded, which means that the boundedness is not necessarily preserved by continuous functions.

Now, we introduce the definition of a totally bounded set and show that the total boundedness is preserved by continuous functions.

Definition 6.4.2. A subset B of a metric space (X, d) is said to be *totally bounded* if, for every $\varepsilon > 0$, there exist $a_1, a_2, \ldots, a_N \in X$ such that

$$B \subseteq \bigcup_{n=1}^{N} B_\varepsilon(a_n).$$

Example 6.4.1. (1) The interval $[0, 1]$ is totally bounded in $\mathbb{R}$ because it can be covered by the balls $B_\varepsilon(n\varepsilon)$ for every $n = 0, 1, 2 \ldots, N$, where $\frac{1}{\varepsilon} - 1 < N \le \frac{1}{\varepsilon}$;

(2) Every bounded set need not be a totally bounded set. For example, in the discrete metric space (X, d), any subset is bounded, but only finite subsets are totally bounded (take $\varepsilon < 1$);

(3) Any subset of a totally bounded set is also totally bounded;

(4) A totally bounded set is bounded because it is a subset of a finite number of bounded balls.

Theorem 6.4.4. *A finite union of totally bounded sets is totally bounded.*

Proof. We can prove this theorem by using the definition of a totally bounded set.

Theorem 6.4.5. *Let (X, d) and (Y, d^*) be metric spaces and $f : X \to Y$ be a uniformly continuous on X. If B is a totally bounded subset of X, then $f(B)$ is a totally bounded subset in Y.*

Proof. Since $f : B \to Y$ is uniformly continuous on B, for every $\varepsilon > 0$, there exists $\delta > 0$ such that, for every $x \in B$,

$$f(B_\delta(x)) \subseteq B_\varepsilon(f(x)).$$

Since B is totally bounded, B is covered by a finite number of balls, that is,

$$B \subseteq \bigcup_{n=1}^{N} B_\delta(x_n).$$

Therefore, we have

$$f(B) \subseteq \bigcup_{n=1}^{N} f(B_\delta(x_n)) \subseteq \bigcup_{n=1}^{N} B_\varepsilon(f(x_n)),$$

which implies that $f(B)$ is totally bounded in Y. This completes the proof.

Now, we give the definition of a compact set and some properties of compact sets.

Definition 6.4.3. Let (X, d) be a metric space and Λ be an index set.

(1) A collection $\tau = \{O_\alpha\}_{\alpha \in \Lambda}$ of open subsets of X is called a *open cover* of X if

$$X \subseteq \bigcup_{\alpha \in \Lambda} O_\alpha;$$

(2) X is said to be *compact* if there exists a finite sub-collection $\{O_{\alpha_1}, O_{\alpha_1}, \ldots, O_{\alpha_n}\}$ of the open cover τ of X such that

$$X \subseteq \bigcup_{i=1}^{n} O_{\alpha_i};$$

(3) In this case, a finite sub-collection $\{O_{\alpha_1}, O_{\alpha_1}, \ldots, O_{\alpha_n}\}$ of the open cover τ of X is called a *finite sub-cover* of X;

(4) A non-empty subset K of a metric space (x, d) is said to be *compact* if it is a compact metric space with the metric induced by the metric d.

Thus, we can say that X is *compact* if a collection $\tau = \{O_\alpha\}_{\alpha \in \Lambda}$ of open subsets of X has a finite sub-cover of X.

Example 6.4.2. (1) A finite subset $K = \{x_1, x_2, \ldots, x_n\}$ of a metric space (X, d) is compact;

(2) Let $K = [0, \infty)$ be a subset of the usual metric space $\mathbb{R}$. Then, K is not compact;

(3) The set $\mathbb{Z}$ of integers with the usual metric $d(m, n) = |m - n|$ for every $m, n \in \mathbb{Z}$ is not compact;

(4) The Euclidean metric space $\mathbb{R}$ is not compact;

(5) Let $K = (0, 1)$ be a subset of the Euclidean metric space $\mathbb{R}$. Then, K is not compact. But $[0, 1]$ is compact.

Solution. (1) Let $\tau = \{O_\alpha\}_{\alpha \in \Lambda}$ of be any open cover of Y. Then, each x_i $(i = 1, 2, \ldots, n)$ is contained in some set O_{α_i} in τ. Then, we have

$$K = \{x_1, x_2, \ldots, x_n\} \subseteq O_{\alpha_1} \bigcup O_{\alpha_2} \bigcup \cdots \bigcup O_{\alpha_n}.$$

That is, the collection $\tau = \{O_\alpha\}_{\alpha \in \Lambda}$ has a finite sub-cover of K. Therefore, a finite set $K = \{x_1, x_2, \ldots, x_n\}$ is compact.

(2) Let $\tau = \{O_n := (-1, n) : n \geq 1\}$. Then, we have

$$K = [0, \infty) \subseteq \bigcup_{n=1}^{\infty} O_n,$$

and so $\tau = \{O_n := (-1, n) : n \geq 1\}$ is an open cover of K. Let $\{O_{n_1}, O_{n_2}, \ldots, O_{n_k}\}$ be any sub-collection of τ. If $m := \sup\{n_1, n_2, \ldots, n_k\}$, then we have

$$O_{n_1} \bigcup O_{n_2} \bigcup \cdots \bigcup O_{n_k} = G_m = (-1, m) \subset K = [0, \infty),$$

and so τ does not have a finite sub-cover of $K = [0, \infty)$. Therefore, $K = [0, \infty)$ is not compact.

(3) Since $\{n\} = \mathbb{Z} \bigcap (n - \frac{1}{2}, n + \frac{1}{2})$ for every $n \in \mathbb{Z}$, the collection $\tau = \{\{n\} : n \in \mathbb{Z}\}$ is an open cover of $\mathbb{Z}$. However, τ does not have a finite sub-cover of $\mathbb{Z}$. Therefore, $\mathbb{Z}$ is not compact.

(4) Let $\tau = \{O_n := (-n, n) : n \geq 1\}$ be a collection of open intervals in $\mathbb{R}$. Then, we have

$$\mathbb{R} = (-\infty, \infty) = \bigcup_{n=1}^{\infty} O_n,$$

and so $\tau = \{O_n := (-n, n) : n \geq 1\}$ is an open cover of $\mathbb{R}$. Let $\{O_{n_i} := (-n_i, n_i) : 1 \leq i \leq k\}$ be any finite open sub-collection of τ and $m := \sup\{n_1, n_2, \ldots, n_k\}$. Then, we have

$$m \notin \bigcup_{i=1}^{k} O_{n_i},$$

and so the open sub-collection $\{O_{n_i} := (-n_i, n_i) : 1 \leq i \leq k\}$ is not an open sub-cover of $\mathbb{R}$. Therefore, $\mathbb{R}$ is not compact.

(5) Let $\tau = \{O_n := (\frac{1}{n}, 1) : n \geq 1\}$ be a collection of open intervals in $\mathbb{R}$. Then, we have

$$K = (0, 1) = \bigcup_{n=1}^{\infty} O_n,$$

and so $\tau = \{O_n := (\frac{1}{n}, 1) : n \geq 1\}$ is an open cover of $K = (0, 1)$. Let $\{O_{n_1}, O_{n_2}, \ldots, O_{n_k}\}$ be any sub-collection of τ. If $m := \sup\{n_1, n_2, \ldots, n_k\}$, then we have

$$O_{n_1} \bigcup O_{n_2} \bigcup \cdots \bigcup O_{n_k} = G_m = \left(\frac{1}{m}, 1\right) \subset K = (0, 1),$$

and so τ does not have a finite sub-cover of $K = (0, 1)$. Therefore, $K = (0, 1)$ is not compact.

Theorem 6.4.6. *Let $\{K_\alpha\}_{\alpha \in \Lambda}$ be a collection of compact subsets of a metric space (X, d). Then, we have the following:*

(1) $\bigcap_{\alpha \in \Lambda} K_\alpha$ *is compact;*

(2) *If Λ is finite, then $\bigcup_{\alpha \in \Lambda} K_\alpha$ is compact.*

Proof. By using the definition of a compact set, we can prove (1) and (2).

Theorem 6.4.7. *If K is a closed subset of a compact metric space (X, d), then K is compact.*

Proof. Let K be closed subset of a compact metric space X and $\tau = \{O_\alpha\}_{\alpha \in \Lambda}$ be an open cover of K, that is,

$$K \subseteq \bigcup_{\alpha \in \Lambda} O_\alpha.$$

Then, since X is compact, from

$$X \subseteq K \bigcup (X - K) \subseteq \bigcup_{\alpha \in \Lambda} O_\alpha \bigcup (X - K),$$

we have

$$X \subseteq \bigcup_{i=1}^{n} O_{\alpha_i} \bigcup (X - K),$$

and so, since $K^c \cap K = \emptyset$,

$$K \subseteq \bigcup_{i=1}^{n} O_{\alpha_i},$$

that is, the finite collection $\{O_{\alpha_1}, O_{\alpha_2}, \ldots, O_{\alpha_n}\}$ is a finite open cover of K. Therefore, K is compact. This completes the proof.

Theorem 6.4.8. *Let (X, d) and (Y, d^*) be metric spaces and $f : X \to Y$ be a continuous mapping. If K is a compact subset of X, then $f(K)$ is compact in Y.*

Proof. Let $\tau = \{O_\alpha\}_{\alpha \in \Lambda}$ be an open cover of $f(K)$. Then, since O_α is open in Y and $f : X \to Y$ is a continuous mapping on X, for every $\alpha \in \Lambda$, $f^{-1}(O_\alpha)$ is open in X. Also, the collection $\{K \cap f^{-1}(O_\alpha)\}$ is an open cover of K. Since K is compact, we have a finite open sub-cover $\{O_{\alpha_1}, O_{\alpha_2}, \ldots, O_{\alpha_n}\}$ of τ such that

$$K \subseteq \bigcup_{i=1}^{n} \left(K \cap f^{-1}(O_{\alpha_i}) \right) = K \cap \left(\bigcup_{i=1}^{n} f^{-1}(O_{\alpha_i}) \right) = K \cap f^{-1} \left(\bigcup_{i=1}^{n} O_{\alpha_i} \right),$$

which implies

$$K \subseteq f^{-1} \left(\bigcup_{i=1}^{n} O_{\alpha_i} \right)$$

and so

$$f(K) \subseteq \bigcup_{i=1}^{n} O_{\alpha_i}.$$

Therefore, the collection $\{O_{\alpha_1}, O_{\alpha_2}, \ldots, O_{\alpha_n}\}$ is a finite sub-cover of τ, and so $f(K)$ is compact. This completes the proof.

Theorem 6.4.9. *Let (X, d) and (Y, d^*) be metric spaces and $f : X \to Y$ be a continuous mapping on X. If X is compact and F is a closed subset of X, then $f(F)$ is closed in Y.*

Proof. Let F be a closed subset of X. Since X is compact, by Theorem 6.4.7, F is compact. Therefore, since $f : X \to Y$ is continuous on X, by Theorem 6.4.8, $f(F)$ is compact, and so $f(F)$ is closed. This completes the proof.

Theorem 6.4.10. *Let (X, d) and (Y, d^*) be metric spaces and $f : X \to Y$ be a continuous mapping on X. If X is compact and f is bijective, that is, onto and one-to-one, then f^{-1} is continuous on Y.*

Proof. Since f is bijective on X, f^{-1} is also bijective on Y. Let F be a closed subset of X. Then, we have

$$(f^{-1})^{-1}(F) = f(F)$$

and, by Theorem 6.4.9, $f(F)$ is closed in Y. Thus, $(f^{-1})^{-1}(F)$ is closed, and so f^{-1} is continuous on Y. This completes the proof.

6.5   The Finite Intersection Property

In this section, we give some properties of the finite intersection property and some relations between compact sets and the finite intersection properties in metric spaces.

Definition 6.5.1. We say that a collection $\tau = \{U_1, U_2, \ldots, U_n, \ldots\}$ of subsets of a metric space (X, d) have the *finite intersection property* if every finite sub-collection $\{U_1, U_2, \ldots, U_n\}$ of τ has the non-empty intersection, that is, for every finite sub-collection $\{U_1, U_2, \ldots, U_n\}$ of τ,

$$\bigcap_{i=1}^{n} U_n \neq \emptyset.$$

Example 6.5.1. (1) The collection $\tau = \{[-\frac{1}{n}, \frac{1}{n}] : n \geq 1\}$ of all closed intervals has the finite intersection property;

(2) Similarly, the collection $\tau = \{(0, \frac{1}{n}) : n \geq 1\}$ of open intervals has the finite intersection property.

Theorem 6.5.1. *Let (X, d) be a metric space. Then, the following are equivalent:*

(1) X *is compact;*

(2) *If $\tau = \{F_\alpha : \alpha \in \Lambda\}$ is the collection of closed subsets of X with the finite intersection property, then*

$$\bigcap_{\alpha \in \Lambda} F_\alpha \neq \emptyset.$$

Proof. (1) $\Longrightarrow$ (2) Let X be compact, and suppose that the collection $\tau = \{F_\alpha : \alpha \in \Lambda\}$ of closed subsets of X has the finite intersection property. Assume that

$$\bigcap_{\alpha \in \Lambda} F_\alpha = \emptyset.$$

Then, we have

$$\bigcup_{\alpha \in \Lambda} (X - F_\alpha) = X - \left(\bigcap_{\alpha \in \Lambda} F_\alpha \right) = X,$$

which implies that the collection $\{X - F_\alpha\}_{\alpha \in \Lambda}$ is an open cover of X, where $X - F_\alpha$ is open for each $\alpha \in \Lambda$ since each F_α is closed. Since X is compact, this collection has a finite open sub-cover $\{X - F_{\alpha_i} : i = 1, 2, \ldots, n\}$ of X and so

$$\bigcup_{i=1}^{n} (X - F_{\alpha_i}) = X - \left(\bigcap_{i=1}^{n} F_{\alpha_i} \right) = X.$$

Thus, we have

$$\bigcap_{i=1}^{n} F_{\alpha_i} = \emptyset,$$

which shows that the collection $\{X - F_{\alpha_i} : i = 1, 2, \ldots, n\}$ does not have the finite intersection property. This is a contradiction. Therefore, we have

$$\bigcap_{\alpha \in \Lambda} F_\alpha \neq \emptyset.$$

$(2) \implies (1)$ Assume that X is not compact. Then, there exists an open cover $\{O_\alpha : \alpha \in \Lambda\}$ of open sets in X which does not have a finite sub-cover of X. This implies that, for every finite sub-collection $\{O_{\alpha_i} : i = 1, 2, \ldots, n\}$ of $\{O_\alpha : \alpha \in \Lambda\}$, we have

$$\bigcup_{i=1}^{n} O_{\alpha_i} \neq X \implies X - \left(\bigcup_{i=1}^{n} O_{\alpha_i}\right) \neq \emptyset \implies \bigcup_{i=1}^{n} (X - O_{\alpha_i}) \neq \emptyset.$$

For each $\alpha \in \Lambda$, let $F_{\alpha \in \Lambda} = X - O_\alpha$. Then, for every $\alpha \in \Lambda$, F_α is closed. Since

$$\bigcap_{i=1}^{n} F_{\alpha_i} = \bigcap_{i=1}^{n} (X - O_{\alpha_i}) \neq \emptyset,$$

$\{F_\alpha : \alpha \in \Lambda\}$ is a collection of closed sets with the finite intersection property. But we have

$$\bigcap_{i=1}^{n} F_{\alpha_i} \neq \emptyset.$$

On the other hand, since $\{O_\alpha : \alpha \in \Lambda\}$ is an open cover of X, we have

$$X = \bigcup_{\alpha \in \Lambda} O_\alpha$$

and

$$\bigcap_{\alpha \in \Lambda} F_{\alpha_i} = \bigcap_{\alpha \in \Lambda} (X - O_\alpha) = X - \left(\bigcup_{\alpha \in \Lambda} O_\alpha\right) = \emptyset,$$

which is a contradiction. Therefore, X is compact. This completes the proof.

By using Theorem 6.5.1, we prove the following:

Theorem 6.5.2. Let $\{F_n : n \geq 1\}$ be a family of non-empty closed subsets of a compact metric space (X, d) with

$$F_{n+1} \subset F_n \quad \text{for every} \ n \geq 1.$$

Then, $\bigcap_{n \geq 1} F_n \neq \emptyset$.

Proof. Since $F_{n+1} \subset F_n$ for every $n \geq 1$, it follows that the family $\{F_n : n \geq 1\}$ has the finite intersection property. Therefore, by Theorem 6.5.1, we have

$$\bigcap_{n \geq 1} F_n \neq \emptyset.$$

This completes the proof.

6.6   The Heine–Borel Theorem

In this section, we introduce the Bolzano–Weierstrass Theorem and the Heine–Borel Theorem in $\mathbb{R}$ and give some relations among compact spaces, the Bolzano–Weierstrass Theorem, and the Heine–Borel Theorem in metric spaces.

The following theorem is well known in $\mathbb{R}$:

Theorem 6.6.1. (The Bolzano–Weierstrass Theorem) *Let C be a closed and bounded subset of $\mathbb{R}$. Then, every infinite subset of C has a limit point in C.*

Note that the Bolzano–Weierstrass Theorem in $\mathbb{R}$ is not true in metric spaces. For example, consider a metric space (C, d), where $C = (0, 1]$ and d is the usual metric. Then, C is closed and bounded. Let $\{1, \frac{1}{2}, \frac{1}{3}, \ldots, \frac{1}{n}, \ldots\}$ be an infinite subset of C. Then, we know that C has only one limit point 0, but $0 \notin C$.

From this property of the Bolzano–Weierstrass Theorem, we can consider the following definition:

Definition 6.6.1. We say that a metric space (X, d) has the *Bolzano–Weierstrass property* if every infinite subset of X has a limit point.

Example 6.6.1. By Theorem 6.6.1, every closed and bounded subset $[a, b]$ of $\mathbb{R}$ has the Bolzano–Weierstrass property.

Theorem 6.6.2. *If a metric space (X, d) is compact, then X has the Bolzano–Weierstrass property.*

Proof. Let C be an infinite subset of a compact metric space (X, d), and assume that C has no any limit point. Then, for any point $x \in X$, there exists an open ball $B_{r_x}(x) = \{y \in X : d(x, y) < r_x\}$ such that

$$B_{r_x}(x) \bigcap C = \begin{cases} \emptyset, & \text{if } x \notin C, \\ \{x\}, & \text{if } x \in C. \end{cases} \qquad (6.6.1)$$

Thus, the collection $\tau = \{B_{r_x}(x) : x \in X\}$ is an open cover of a metric space X. Since X is compact, the collection τ has a finite sub-cover $\{B_{r_{x_i}}(x_i) : x_i \in X, i = 1, 2, \ldots, n\}$ of X, that is,

$$X \subseteq \bigcup_{i=1}^{n} B_{r_{x_i}}(x_i).$$

Therefore, by (6.6.1), we have

$$C = C \bigcap X$$

$$\subseteq C \bigcap \left(\bigcup_{i=1}^{n} B_{r_{x_i}}(x_i) \right)$$

$$= \bigcup_{i=1}^{n} \left(C \bigcap B_{r_{x_i}}(x_i) \right)$$

$$\subseteq \bigcup_{i=1}^{n} \{x_i\}$$

$$= \{x_1, x_2, \ldots, x_n\}$$

and so C is a finite set, which is a contradiction. Hence, C has a limit point. This completes the proof.

Definition 6.6.2. A metric space (X, d) is said to be *sequentially compact* if every sequence $\{x_n\}$ in X has a convergent subsequence $\{x_{n_k}\}$ of $\{x_n\}$.

Example 6.6.2. (1) Every finite subset of a metric space (X, d) is sequentially compact;

(2) An open interval $(0, 1)$ in $(\mathbb{R}, d)$ with the usual metric d is not sequentially compact.

Now, we give some relations between compact spaces and the Bolzano–Weierstrass property as follows:

Theorem 6.6.3. *Let (X, d) be a metric space. Then, the following are equivalent:*

(1) *X is sequentially compact;*

(2) *X has the Bolzano–Weierstrass property.*

Proof. (1) $\Longrightarrow$ (2) Suppose that X is sequentially compact. Let C be an infinite subset of X. Then, we can construct a sequence $\{x_n\}$ since C is an infinite subset of X. Since X is sequentially compact, the sequence $\{x_n\}$ has a convergent subsequence $\{x_{n_k}\}$, say $x_{n_k} \to x \in X$ as $k \to \infty$.

Now, we prove that the point $x \in X$ is a limit point of C. In fact, since $x_{n_k} \to x \in X$ as $k \to \infty$, for every $\varepsilon > 0$, there exists a positive integer N such that, for every $k \geq N$,

$$|x_{n_k} - x| < \varepsilon,$$

that is,

$$x_{n_k} \in B_\varepsilon(x) \text{ for every } k \geq N.$$

Since $\{x_n\}$ is a sequence to be constructed from infinitely many distinct points in C, we have $x_{n_k} \neq x$ for every $k \geq N$. Thus, by the definition of a limit point, the point x is a limit point of C, and so X has the Bolzano–Weierstrass property.

(2) $\Longrightarrow$ (1) Assume that X has the Bolzano–Weierstrass property. Then, every infinite subset C of X has a limit point. Let $\{x_n\}$ be a sequence in X.

Now, we prove that the sequence $\{x_n\}$ has a convergent subsequence. In fact, let $C = \{x_1, x_2, \ldots, x_n, \ldots\}$ be a set of points $x_1, x_2, \ldots, x_n, \ldots$.

Now, we have two cases as follows:

Case 1. If C is a finite set, then we can consider a sequence $\{x_n\}$ as follows:

$$\{x_n\} = \{x_1, x_2, \ldots, x_n, x, x, \ldots\}.$$

Thus, we have a convergent subsequence $\{x, x, x, \ldots\}$ of $\{x_n\}$ converging to the point x.

Case 2. If C is an infinite set, then the sequence $\{x_n\}$ has infinitely many distinct points. Since X has a limit point, say x, the open ball $S_1(x)$ has infinitely many points from C, that is,

$$S_1(x) \bigcap (C - \{x\}) \neq \emptyset.$$

So, we can choose a point $x_{n_1} \in S_1(x) \bigcap (C - \{x\})$. Again, we have

$$S_{\frac{1}{2}}(x) \bigcap (C - \{x\}) \neq \emptyset,$$

and so we can choose a point $x_{n_2} \in S_{\frac{1}{2}}(x) \bigcap (C-\{x\})$ with $n_2 > n_1$. Similarly, we can have a subsequence $\{x_{n_k}\}$ of the sequence $\{x_n\}$ with $n_k > n_{k-1} > \cdots > n_2 > n_1$ and

$$x_{n_k} \in S_{\frac{1}{k}}(x) \bigcap (C - \{x\}),$$

that is,

$$d(x_{n_k}, x) < \frac{1}{k}.$$

Therefore, if $k \to \infty$, then we have $x_{n_k} \to x$, and so the sequence $\{x_n\}$ has a convergent subsequence $\{x_{n_k}\}$, which implies that X is sequentially compact. This completes the proof.

Corollary 6.6.1. *If a metric space (X, d) is compact, then X is sequentially compact.*

The following Heine–Borel Theorem is well known in $\mathbb{R}$.

Theorem 6.6.4. (The Heine–Borel Theorem) *Let C be a subset of $\mathbb{R}$. Then, the following are equivalent:*

(1) C is closed and bounded;

(2) C is compact.

Now, by using Corollary 6.6.1, we prove the Heine–Borel Theorem in metric spaces as follows.

Theorem 6.6.5. *If C is a compact subset of a metric space (X, d), then C is closed and bounded.*

Proof. Let C be a compact subset of a metric space (X, d).

(a) If C is a finite set, then clearly every finite set is closed and bounded;

(b) Let C be an infinite set and $x \in \overline{C}$, where $\overline{C}$ is the closure of the set C. Then, there exists a sequence $\{x_n\}$ in C such that $x_n \to x$ as $n \to \infty$. Since C is compact, by Corollary 6.6.1, C is sequentially compact, and so the sequence $\{x_n\}$ has a convergent subsequence $\{x_{n_k}\}$. Let $x_{n_k} \to x \in C$. Then, we have $\overline{C} \subset C$ and so $\overline{C} = C$, that is, C is closed.

Next, we prove that C is bounded. In fact, suppose that C is not bounded. Then, there exist $x, y \in C$ such that

$$d(x, y) > M \quad \text{for some} \quad M > 0.$$

Now, consider the collection $\tau = \{B_1(x) : x \in C\}$ of open balls with center x and radius 1. Then, we have

$$C \subseteq \bigcup_{x \in C} S_1(x),$$

which implies that the collection $\tau = \{B_1(x) : x \in C\}$ is an open cover of C. Since C is compact, the collection τ has a finite sub-cover $\{B_1(x_i) : i = 1, 2, \ldots, n\}$ of C, that is,

$$C \subseteq \bigcup_{i=1}^{n} B_1(x_i).$$

Let $k = \max\{d(x_i, x_j) : i, j = 1, 2, \ldots, n, i \neq j\}$ and $M = k + 2$. Then, we have

$$d(x, y) > k + 2 \quad \text{for every} \quad x, y \in C \tag{6.6.2}$$

and there exist $x_i, x_j \in C$ such that

$$x \in B_1(x_i), \quad y \in B_1(x_j).$$

So, we have

$$d(x, y) \leq d(x, x_i) + d(x_i, x_j) + d(x_j, x) < k + 2,$$

which is a contradiction to (6.6.2). Therefore, C is bounded. This completes the proof.

Note that the converse of Theorem 6.6.5 is not true always. For example, consider an infinite subset C of a discrete metric space (x, d). Then, C is closed and bounded. But the collection $\{\{x\} : x \in C\}$ is an open cover of C, but the collection $\{\{x\} : x \in C\}$ has no finite sub-cover of C, and so C is not compact.

Theorem 6.6.6. *Let K be a compact subset of a metric space (X, d). If F is a closed subset of K, then F is compact.*

Proof. Let the collection $\{O_\lambda : \lambda \in \Lambda\}$ be an open cover of F. Then, the open collection $\{O_\lambda : \lambda \in \Lambda\} \bigcup \{F^c\}$ covers K and also F because $F \subset K$. Since K is compact, we can have a finite sub-cover of the collection $\{O_\lambda : \lambda \in \Lambda\} \bigcup \{F^c\}$ for K.

If F^c is an element of this finite sub-cover, then we simply remove F^c to get a finite sub-cover of F by the collection $\{O_\lambda : \lambda \in \Lambda\}$. This completes the proof.

Theorem 6.6.7. *If K is a compact subset of a metric space (x, d) and F is a closed subset of X, then $F \bigcap K$ is compact.*

Proof. By using Theorems 6.6.5 and 6.6.6, we can complete the proof.

6.7 Exercises

Excercise 6.1 Let (X, d_1) and (Y, d_2) be metric spaces and $f : X \to Y$ be a continuous function. Prove that, if C is a connected subset of X, then $f(C)$ is a connected subset of Y.

Hint. Let C be a connected subset of X. If possible, suppose that $f(C)$ is not connected, that is, disconnected. Then, one can write $f(C) = A \bigcup B$, where A and B are non-empty and disjoint open subsets of Y such that $f(C) \bigcap A$ and $f(C) \bigcap B$ are non-empty. Then, one should arrive at a contradiction to the assumption that C is connected from the following inclusion and the continuity of f:

$$C \subseteq f^{-1}(A \bigcup B) = f^{-1}(A) \bigcup f^{-1}(B).$$

Excercise 6.2 Prove that the connected subsets of the set $\mathbb{Q}$ of rational numbers are the only one point sets.

Hint. Let X be a connected subspace of $\mathbb{Q}$ containing two points, say x and y. Then, one can find an irrational number α lying between x and y. Also,

$$X = \left[X \bigcap (-\infty, \alpha) \right] \bigcup \left[X \bigcap (\alpha, \infty) \right]$$

and

$$\left[X \bigcap (-\infty, \alpha) \right] \bigcap \left[X \bigcap (\alpha, \infty) \right] = \emptyset.$$

One should now arrive at a contradiction to the hypothesis that X is connected.

Excercise 6.3 Prove that a subset $A = \{x \in \mathbb{R} : |x| > 0\}$ of $\mathbb{R}$ is disconnected.

Hint. $A = [A \bigcap (-\infty, 0)] \bigcup [A \bigcap (0, \infty)]$.

Excercise 6.4 Let M be a subset of a metric space (X, d). Prove that if every pair of points in M lies in a connected subset of M, then M is connected.

Hint. Suppose that M is not connected. Then, there exist non-empty separated sets A and B such that $M = A \cup B$. Let $x \in A$ and $y \in B$. Then, an application of the hypothesis for the pair of points x and y should lead one to a contradiction.

Excercise 6.5 Prove that a connected subset A of a metric space (X, d), which is both open and closed, is a component of X.

Hint. Let C be a component of X. Then, every connected subset of X is contained in C, that is, $A \subseteq C$. Now, one can claim that $A = C$.

If possible, suppose that $A \neq C$. Now, note that

$$\left[(C \cap A) \right] \cap \left[C \cap (X - A) \right] = C \cap \left[A \cap (X - A) \right] = C \cap \emptyset = \emptyset$$

and

$$\left[(C \cap A) \right] \cup \left[C \cap (X - A) \right] = C \cap \left[A \cup (X - A) \right] = C \cap X = C.$$

Excercise 6.6 Prove the following:

(1) If A_i, B are connected sets and, for all $i \geq 1$, $A_i \cap B \neq \emptyset$, then $B \cup \left(\bigcup_{i=1}^{\infty} A_i \right)$ is connected;

(2) If A_n are connected for each $n \geq 1$ and $A_n \cap A_{n+1} \neq \emptyset$, then $\bigcup_{n=1}^{\infty} A_n$ is connected.

Excercise 6.7 Prove that a finite union of compact sets is compact.

Hint. Suppose that the open sets O_i cover the finite union of the compact sets $K_1 \cup K_2 \cup \cdots \cup K_N$. Then, they cover each $K_n (n = 1, 2, \ldots, N)$. So, they have a finite cover of each K_n. It is easy to see that

$$K_n \subseteq \bigcup_{k=1}^{N} O_{i_k}.$$

Then, for each $n = 1, 2, \ldots, N$, the collection $\{ O_{i_k} : k = 1, 2, \ldots, N \}$ is finite and together covers all the K_ns.

Excercise 6.8 Which of the following sets are compact? Justify your answers.

(1) $A = [0, 1] \cup [2, 3] \subset \mathbb{R}$;

(2) $A = \{ x \in \mathbb{R} : x \geq 0 \} \subset \mathbb{R}$;

(3) $A = \mathbb{Q} \cap [0, 1] = \{ x \in \mathbb{R} : 0 \leq x \leq 1 \text{ and } x \text{ is irrational} \}$;

(4) $A = \{ 1, \frac{1}{2}, \frac{1}{3}, \ldots, \frac{1}{n}, \ldots \} \cup \{ 0 \}$.

Ans. (1) Yes;
     (2) No;
     (3) No;
     (4) Yes.

Excercise 6.9 Let (X, d) be a metric space and $\{x_n\}$ be a sequence in X such that $x_n \to x$ as $n \to \infty$. Prove that a subset $\{x\} \bigcup \{x_n : n \geq 1\}$ of X is compact.

Excercise 6.10 Let A be a compact subset of a metric space (X, d). Prove that, for any subset B of X, there exists a point $x \in A$ such that

$$\rho(x, B) = \rho(A, B),$$

where $\rho(x, B) = \inf_{y \in B} d(x, y)$ and $\rho(A, B) = \inf\{d(x, y) : x \in A, y \in B\}$.

7

Differentiation

In this chapter, we deal with the definition of the derivative, the differentiable functions and their properties, the chain rule, derivative of inverse functions, Rolle's Theorem, Lagrange's Mean Value Theorem, Intermediate Value Theorem for derivative, Cauchy's Mean Value Theorem, indeterminate forms, L'Hospital Rule, Taylor's Theorem, Taylor series, Maclaurin series, and the local minimum and maximum. Several solved examples and chapter-end exercises are also incorporated.

7.1   The Derivative

First, we give the definition of the derivative of a function.

Definition 7.1.1. Let E be a subset of $\mathbb{R}$ and $f : E \to \mathbb{R}$ be a real-valued function on E containing a point x_0.

(1) We say that f is *differentiable* at x_0 if the limit

$$f'(x_0) = \lim_{x \to x_0} \frac{f(x) - f(x_0)}{x - x_0}$$

exists and is finite or, equivalently, for any $\epsilon > 0$, there exists $\delta > 0$ such that, for any $x \in E$ with $0 < |x - x_0| < \delta$,

$$\left| \frac{f(x) - f(x_0)}{x - x_0} - f'(x_0) \right| < \epsilon;$$

(2) $f'(x_0)$ is called the *derivative* of f at x_0;

(3) If f is not differentiable at every point $x \in E$, then f is said to be differentiable on E;

(4) If the derivative $f'(x_0)$ does not exist, then f is not differentiable at $x = x_0$.

Example 7.1.1. Define a function $f : (0, \infty) \to \mathbb{R}$ by $f(x) = \sqrt{x}$ for all $x \in (0, \infty)$. Show that f is differentiable at every point in $(0, \infty)$.

Solution. For every point $x_0 \in (0, \infty)$, we can see the following:

$$
\begin{aligned}
f'(x_0) &= \lim_{x \to x_0} \frac{\sqrt{x} - \sqrt{x_0}}{x - x_0} \\
&= \lim_{x \to x_0} \frac{x - x_0}{(x - x_0)(\sqrt{x} + \sqrt{x_0})} \\
&= \lim_{x \to x_0} \frac{1}{\sqrt{x} + \sqrt{x_0}} \\
&= \frac{1}{2\sqrt{x_0}},
\end{aligned}
$$

which implies that f is differentiable at $x = x_0$ and so, since $x_0 \in (0, \infty)$ is arbitrary, f is differentiable at every point $x \in (0, \infty)$.

Example 7.1.2. Define a function $f : (0, \infty) \to \mathbb{R}$ by

$$
f(x) = \begin{cases} x \sin \frac{1}{x}, & \text{if } x \neq 0, \\ 0, & \text{if } x = 0. \end{cases}
$$

Show that f is not differentiable at $x = 0$.

Solution. For $x = 0$, we can see the following:

$$
\lim_{x \to 0} \frac{f(x) - f(0)}{x - 0} = \lim_{x \to 0} \frac{x \sin \frac{1}{x} - 0}{x - 0} = \lim_{x \to 0} \sin \frac{1}{x},
$$

which implies that the limit $\lim_{x \to 0} \sin \frac{1}{x}$ does not exist and so f is not differentiable at $x = 0$.

Example 7.1.3. For every $n \geq 1$, let $f(x) = x^n$ for every $x \in \mathbb{R}$. Show that $f'(x) = nx^{n-1}$ for every $x \geq 1$.

Solution. For every fixed $x_0 \in \mathbb{R}$, observe that

$$
\begin{aligned}
&f(x) - f(x_0) \\
&= x^n - x_0^n \\
&= (x - x_0)(x^{n-1} + x_0 x^{n-2} + x_0^2 x^{n-3} + \cdots + x_0^{n-2} x) + x_0^{n-1})
\end{aligned}
$$

and so

$$
\begin{aligned}
f'(x_0) &= \lim_{x \to x_0} \frac{f(x) - f(x_0)}{x - x_0} \\
&= \lim_{x \to x_0} (x^{n-1} + x_0 x^{n-2} + x_0^2 x^{n-3} + \cdots + x_0^{n-2} x) + x_0^{n-1}) \\
&= x_0^{n-1} + x_0 x_0^{n-2} + x_0^2 x_0^{n-3} + \cdots + x_0^{n-2} x_0 + x_0^{n-1} \\
&= nx_0^{n-1}.
\end{aligned}
$$

Therefore, we have $f'(x_0) = nx_0^{n-1}$ and so, since $x_0 \in \mathbb{R}$ is arbitrary, $f'(x) = nx^{n-1}$ for all $x \in \mathbb{R}$.

Definition 7.1.2. Let E be a subset of $\mathbb{R}$ and $f : E \to \mathbb{R}$ be a real-valued function on E containing a point x_0.

(1) If x_0 is not the right endpoint of E, then f is said to be *right-sided differentiable* at x_0 if

$$f'_+(x_0) = \lim_{x \to x_0^+} \frac{f(x) - f(x_0)}{x - x_0}$$

exists;

(2) If x_0 is not the left endpoint of E, then f is said to be *left-sided differentiable* at x_0 if

$$f'_-(x_0) = \lim_{x \to x_0^-} \frac{f(x) - f(x_0)}{x - x_0}$$

exists;

(3) The numbers $f'_+(x_0)$ and $f'_-(x_0)$ are called the *right derivative* and the *left derivative* of f at x_0, respectively.

From Definitions 7.1.1 and 7.1.2, we have the following.

Theorem 7.1.1. *A function $f : E \to \mathbb{R}$ is differentiable at $x_0 \in E$ if and only if $f'_+(x_0)$, $f'_-(x_0)$ exist and $f'_+(x_0) = f'_-(x_0)$.*

Example 7.1.4. Show that the function $f(x) = |x|$ is differentiable on $\mathbb{R}-\{0\}$.

Solution. Observe that

$$f'(x_0) = \lim_{x \to x_0} \frac{|x| - |x_0|}{x - x_0}$$

$$= \begin{cases} \lim_{x \to x_0} \dfrac{x - x_0}{x - x_0} = 1, & \text{if } x_0 > 0, \\ \lim_{x \to x_0} \dfrac{-(x - x_0)}{x - x_0} = -1, & \text{if } x_0 < 0. \end{cases}$$

Therefore, we have $f'_+(x_0) = 1 \neq -1 = f'_-(x_0)$, and so, by Theorem 7.1.1, $f(x) = |x|$ is not differentiable at $x_0 = 0$.

From Example 7.1.4, we know that even though a function f is continuous at x_0, f is not differentiable at x_0. But we have the following.

Theorem 7.1.2. *If $f : E \to \mathbb{R}$ is differentiable at a point $x_0 \in E$, then f is continuous at x_0.*

Proof. Now, we prove $\lim_{x \to x_0} f(x) = f(x_0)$. Since f is differentiable at x_0, we have

$$f'(x_0) = \lim_{x \to x_0} \frac{f(x) - f(x_0)}{x - x_0}.$$

For every $x, x_0 \in E$ with $x \neq x_0$, we have

$$f(x) - f(x_0) = \frac{f(x) - f(x_0)}{x - x_0} \cdot (x - x_0)$$

and so

$$\begin{aligned}
\lim_{x \to x_0} (f(x) - f(x_0)) &= \lim_{x \to x_0} \left[\frac{f(x) - f(x_0)}{x - x_0} \cdot (x - x_0) \right] \\
&= \lim_{x \to x_0} \left[\frac{f(x) - f(x_0)}{x - x_0} \right] \cdot \lim_{x \to x_0} (x - x_0) \\
&= f'(x_0) \cdot 0 \\
&= 0.
\end{aligned}$$

Therefore, $\lim_{x \to x_0} f(x) = f(x_0)$, that is, f is continuous at x_0. This completes the proof.

7.2   The Differential Calculus

In this section, we give some algebraic operations and fundamental properties of differentiable functions.

Theorem 7.2.1. *Let $f, g : E \to \mathbb{R}$ be differentiable at point $x_0 \in E$. Then, the functions $c \cdot f$ (c is a constant), $f + g$, fg, and $\frac{f}{g}$ ($g(x) \neq 0$) are also differentiable at x_0 and the following are satisfied:*

(1) $(cf)'(x_0) = c \cdot f'(x_0)$;

(2) $(f + g)'(x_0) = f'(x_0) + g'(x_0)$;

(3) $(fg)'(x_0) = f(x_0)g'(x_0) + f'(x_0)g(x_0)$;

(4) $\left(\frac{f}{g} \right)'(x_0) = \frac{g(x_0)f'(x_0) - f(x_0)g'(x_0)}{g^2(x_0)}$ *if $g(x_0) \neq 0$.*

Proof. (1) By the definition of $c \cdot f$, we have $(c \cdot f)(x) = c \cdot f(x)$ for every $x \in E$ and so

$$(c \cdot f)'(x_0) = \lim_{x \to x_0} \frac{(c \cdot f)(x) - (c \cdot f)(x_0)}{x - x_0} = \lim_{x \to x_0} c \cdot \frac{f(x) - f(a)}{x - x_0} = c \cdot f'(x_0).$$

(2) Note that

$$\frac{(f + g)(x) - (f + g)(x_0)}{x - x_0} = \frac{f(x) - f(x_0)}{x - x_0} + \frac{g(x) - g(x_0)}{x - x_0}.$$

Thus, taking the limit as $x \to x_0$, we have the following:

$$(f+g)'(x_0) = \lim_{x \to x_0} \frac{(f+g)(x) - (f+g)(x_0)}{x - x_0}$$

$$= \lim_{x \to x_0} \frac{f(x) - f(x_0)}{x - x_0} + \lim_{x \to x_0} \frac{g(x) - g(x_0)}{x - x_0}$$

$$= f'(x_0) + g'(x_0).$$

Therefore, it follows that $(f+g)'(x_0) = f'(x_0) + g'(x_0)$.

(3) Observe that

$$\frac{(fg)(x) - (fg)(x_0)}{x - x_0} = f(x)\frac{g(x) - g(x_0)}{x - x_0} + g(x_0)\frac{f(x) - f(x_0)}{x - x_0}$$

for every $x \in E$ with $x \neq x_0$. Thus, taking the limit as $x \to x_0$, since $\lim_{x \to x_0} f(x) = f(x_0)$, we have the following:

$$(fg)'(a) = \lim_{x \to x_0} \frac{(fg)(x) - (fg)(x_0)}{x - x_0}$$

$$= \lim_{x \to x_0} \left[f(x)\frac{g(x) - g(x_0)}{x - x_0} + g(x_0)\frac{f(x) - f(x_0)}{x - x_0} \right]$$

$$= \lim_{x \to x_0} \left[f(x)\frac{g(x) - g(x_0)}{x - x_0} \right] + \lim_{x \to x_0} g(x_0)\left[\frac{f(x) - f(x_0)}{x - x_0} \right]$$

$$= f(a)g'(a) + g(a)f'(a).$$

Therefore, it follows that $(f+g)'(x_0) = f'(x_0) + g'(x_0)$.

(4) Since $g(x_0) \neq 0$ and g is continuous at x_0, for every $x \in E$, we have the following:

$$\left(\frac{f}{g}\right)(x) - \left(\frac{f}{g}\right)(x_0) = \frac{f(x)}{g(x)} - \frac{f(x_0)}{g(x_0)}$$

$$= \frac{g(x_0)f(x) - f(x_0)g(x)}{g(x)g(x_0)}$$

$$= \frac{g(x_0)f(x) - g(x_0)f(x_0) + g(x_0)f(x_0) - f(x_0)g(x)}{g(x)g(x_0)}$$

and so

$$\frac{\left(\frac{f}{g}\right)(x) - \left(\frac{f}{g}\right)(x_0)}{x - x_0}$$

$$= \left\{ g(x_0)\frac{f(x) - f(x_0)}{x - x_0} - f(x_0)\frac{g(x) - g(x_0)}{x - x_0} \right\}\frac{1}{g(x)g(x_0)}$$

for every $x \in E$ with $x \neq x_0$. Now, taking the limit as $x \to x_0$, we have the following:

$$\left(\frac{f}{g}\right)'(x_0)$$

$$= \lim_{x \to x_0} \left[\frac{\left(\frac{f}{g}\right)(x) - \left(\frac{f}{g}\right)(x_0)}{x - x_0} \right]$$

$$= \lim_{x \to x_0} \left[\left\{ g(x_0)\frac{f(x) - f(x_0)}{x - x_0} - f(x_0)\frac{g(x) - g(x_0)}{x - x_0} \right\} \frac{1}{g(x)g(x_0)} \right]$$

$$= \lim_{x \to x_0} \left[\left\{ g(x_0)\frac{f(x) - f(x_0)}{x - x_0} - f(x_0)\frac{g(x) - g(x_0)}{x - x_0} \right\} \cdot \lim_{x \to x_0} \left[\frac{1}{g(x)g(x_0)} \right] \right]$$

$$= g(x_0)f'(x_0) + f(x_0)g'(x_0) \cdot \frac{1}{g^2(x_0)}$$

$$= \frac{g(x_0)f'(x_0) + f(x_0)g'(x_0)}{g^2(x_0)}$$

for every $x \in E$ with $x \neq x_0$. Therefore, we have

$$\left(\frac{f}{g}\right)'(x_0) = \frac{g(x_0)f'(x_0) - f(x_0)g'(x_0)}{g^2(x_0)}$$

if $g(x_0) \neq 0$. This completes the proof.

Corollary 7.2.2. *For every $k = 1, 2, \ldots, n$, let $f_k : E \to \mathbb{R}$ be differentiable at point $x_0 \in E$. Then, we have the following:*

(1) $f_1 + f_2 + \cdots + f_n$ *is also differentiable at x_0 and*

$$(f_1 + f_2 + \cdots + f_n)'(x_0) = f_1'(x_0) + f_2'(x_0) + \cdots + f_n'(x_0).$$

(2) $f_1 f_2 \cdots f_n$ *is also differentiable at x_0 and*

$$(f_1 f_2 \cdots f_n)'(x_0)$$
$$= f_1'(x_0)f_2(x_0) \cdots f_n(x_0) + f_1(x_0)f_2'(x_0) \cdots f_n(x_0)$$
$$+ f_1(x_0)f_2(x_0) \cdots f_n'(x_0).$$

Theorem 7.2.3. (The Chain Rule) *If $f : A \to B$ is differentiable at $x_0 \in A$ and $g : B \to C$ is differentiable at $f(a) \in B$, then the composite function $g \circ f : A \to C$ is differentiable at $x_0 \in A$ and*

$$(g \circ f)'(x_0) = g'(f(x_0)) \cdot f'(x_0).$$

Proof. It is easy to check that $g \circ f$ is defined on some open interval $E \subset A$ containing x_0. For every $y \in B$ with $y \neq f(x_0)$, let

$$h(y) = \frac{g(y) - g(f(a))}{y - f(a)},$$

and let $h(f(x_0)) = g'(f(x_0))$. Since $\lim_{y \to f(x_0)} h(y) = h(f(x_0))$, the function h is continuous at $f(x_0)$. Since

$$g(y) - g(f(x_0)) = h(y)(y - f(x_0)) \quad \text{for every} \quad y \in B,$$

we have

$$g \circ f(x) - g \circ f(x_0) = h(f(x))(f(x) - f(x_0)) \quad \text{for every} \quad x \in E.$$

Hence, we have

$$\frac{g \circ f(x) - g \circ f(x_0)}{x - x_0} = h(f(x)) \frac{f(x) - f(x_0)}{x - x_0} \tag{7.2.1}$$

for every $x \in E$ with $x \neq x_0$. Since $\lim_{x \to x_0} f(x) = f(x_0)$ and h is continuous at $f(x_0)$, we have

$$\lim_{x \to x_0} h(f(x)) = h(f(x_0)) = g'(f(x_0)).$$

Also, we have

$$\lim_{x \to x_0} \frac{f(x) - f(x_0)}{x - x_0} = f'(x_0).$$

Thus, taking the limit in (7.2.1) as $x \to x_0$, we obtain

$$(g \circ f)'(x_0) = g'(f(x_0)) \cdot f'(x_0).$$

This completes the proof.

Example 7.2.1. Let $f(x) = x^2 + 1$ and $g(x) = \sin x$ for every $x \in \mathbb{R}$. Then, since $h(x) = (g \circ f)(x) = \sin(x^3 + 1)$, by the Chain Rule, we have

$$h'(x) = \cos(x^3 + 1) \cdot (x^3 + 1)' = 3x^2 \cos(x^3 + 1).$$

Now, we consider the derivative of the inverse function. Recall that a function $f : E \to \mathbb{R}$ has the inverse if and only if f is injective, that is, one-to-one. Then, the inverse function f^{-1} has the domain $f(E)$ and is characterized as follows:

$$y = f(x) \iff x = f^{-1}(y).$$

Theorem 7.2.4. (The Derivative of Inverse Functions) *Let $I \subseteq \mathbb{R}$ be an open interval and $f : I \to \mathbb{R}$ be an injective and continuous function on I. If f is differentiable at $x_0 \in I$ and $f'(x_0) \neq 0$, then*

(1) f^{-1} is differentiable at $y_0 := f(x_0)$;

(2) *We have*

$$(f^{-1})'(y_0) = \frac{1}{f'(x_0)} = \frac{1}{f'(f^{-1}(y_0))}.$$

Proof. Let $J := f(I)$ and, for every $y \in J$ with $y \neq y_0$, define a function $H : J \to \mathbb{R}$ by

$$H(y) = \frac{f(f^{-1}(y)) - f(f^{-1}(y_0))}{f^{-1}(y) - f^{-1}(y_0)}. \qquad (7.2.2)$$

Since f^{-1} is injective, we have $f^{-1}(y) \neq f^{-1}(y_0)$ for every $y \in J$ with $y \neq y_0$ and so the function H is well defined on J. Since $y = f(f^{-1}(y_0))$ and $y_0 = f(f^{-1}(y_0))$, it follows from (7.2.2) that

$$H(y) = \frac{y - y_0}{f^{-1}(y) - f^{-1}(y_0)}$$

and so $H(y) \neq 0$ for every $y \in J$ with $y \neq y_0$.

Now, we prove that $\lim_{y \to y_0} H(y) = f'(x_0)$. In fact, since f is differentiable at $x_0 = f^{-1}(y_0) \in I$, for any $\varepsilon > 0$, there exists $\delta > 0$ such that, if $0 < |x - x_0| < \varepsilon$ for every $x \in I$, then

$$\left| \frac{f(x) - f(x_0)}{x - x_0} - f'(x_0) \right| < \varepsilon.$$

But, since f^{-1} is continuous at $y_0 = f(x_0)$, there exists $\gamma > 0$ such that if $|y - y_0| < \gamma$ for every $y \in J$, then $|f^{-1}(y) - f^{-1}(y_0)| < \delta$. Since f^{-1} is injective and $y_0 = f(x_0)$, it follows that if $0 < |y - y_0| < \gamma$ for every $y \in J$, then

$$0 < |f^{-1}(y) - x_0| < \delta.$$

Therefore, we have

$$|H(y) - f^{-1}(x_0)| = \left| \frac{f(f^{-1}(y)) - f(f^{-1}(y_0))}{f^{-1}(y) - f^{-1}(y_0)} - f'(x_0) \right| < \varepsilon$$

when $0 < |y - y_0| < \gamma$ for every $y \in J$. Since $\epsilon > 0$ is arbitrary, we have

$$\lim_{y \to y_0} H(y) = f'(x_0).$$

Since $H(y) \neq 0$ for every $y \in J$ with $y \neq y_0$ and

$$\frac{f^{-1}(y) - f^{-1}(y_0)}{y - y_0} = \frac{1}{H(y)},$$

we have

$$\lim_{y \to y_0} \frac{f^{-1}(y) - f^{-1}(y_0)}{y - y_0} = \lim_{y \to y_0} \frac{1}{H(y)} = \frac{1}{f'(x_0)}.$$

Therefore, $f^{-1}(y_0)$ exists and

$$\left(f^{-1}\right)'(y_0) = \frac{1}{f'(x_0)} = \frac{1}{f'(f^{-1}(y_0))}.$$

This completes the proof.

Note the following:

(1) In Theorem 7.2.4, the hypothesis $f'(x_0) \neq 0$ is essential. In fact, if $f'(x_0) = 0$, then the inverse function f^{-1} is not differentiable at $y_0 = f(x_0)$. If f^{-1} is differentiable at $y_0 = f(x_0)$, since f is the inverse function of f^{-1}, we can apply Theorem 7.2.4 to f^{-1}, that is, f is differentiable at $x_0 = f^{-1}(y_0)$ and $1 = f'(x_0)(f^{-1})'(y_0) = 0$, which is a contradiction. Thus, f^{-1} is not differentiable at $y_0 = f(x_0)$. For example, consider the function $f(x) = x^3$ at $x_0 = 0$;

(2) For a function $y = f(x)$, if $x = f^{-1}(y)$, then we have

$$\left(f^{-1}\right)'(y) = \frac{dx}{dy} = \frac{1}{\frac{dy}{dx}}.$$

Example 7.2.2. Let $f : (0, 2) \to \mathbb{R}$ be a function defined by $f(x) = x^2$ for every $x \in (0, 2)$. Show the following:

(1) Find the inverse of $f(x) = x^2$ for every $x \in (0, 2)$;

(2) Find the derivative of the inverse function.

Solution. (1) The inverse function of $f(x) = x^2$ for every $x \in (0, 2)$ is $f^{-1}(y) = \sqrt{y}$ on $(0, 4)$.

(2) The derivative of the inverse function $f^{-1}(y) = \sqrt{y}$ on $(0, 4)$ is as follows:

$$\left(f^{-1}\right)'(y) = \frac{1}{f'(x)} = \frac{1}{2\sqrt{y}}.$$

Example 7.2.3. Find the derivative of $\sin^{-1} x$ at $x = \frac{1}{2}$.

Solution. Since $\sin x$ is a bijective and differentiable function from $\left[-\frac{\pi}{2}, \frac{\pi}{2}\right]$ into $[-1, 1]$, the inverse of $\sin x$, $\sin^{-1} : [-1, 1] \to \left[-\frac{\pi}{2}, \frac{\pi}{2}\right]$, is also bijective and differentiable. Thus, we have the following:

$$\left[\left(\sin^{-1} y\right)'\right]_{y=\frac{1}{2}} = \frac{1}{\left[(\sin x)'\right]_{x=\frac{\pi}{6}}} = \frac{1}{\cos \frac{\pi}{6}} = \frac{2}{\sqrt{3}}.$$

7.3 Properties of Differentiable Functions

In this section, we give the Mean Value Theorem, which is one of the most useful results in real analysis, and some properties of differentiable functions by using the Mean Value Theorem.

Definition 7.3.1. Let $E \subset \mathbb{R}$ and $f : E \to \mathbb{R}$ be a function.

(1) We say that f has a *local maximum* (resp., *local minimum*) at a point $x_0 \in E$ if there exists $\delta > 0$ such that

$$f(x_0) \geq f(x) \ \text{(resp., } f(x_0) \leq f(x))$$

for every $x \in E$ with $|x - x_0| \leq \delta$;

(2) We say that f has a *local extremum* at a point $x_0 \in E$ if it has a local maximum or a local minimum at x_0;

(3) We say that f has a *global maximum* (resp., *global minimum*) at a point $x_0 \in E$ if

$$f(x_0) \geq f(x) \ \text{(resp., } f(x_0) \leq f(x)) \ \text{for every} \ x \in E.$$

Theorem 7.3.1. (Fermat's Theorem) *Let I be an open interval and $f : I \to \mathbb{R}$ be a function defined on an open interval I containing x_0. If f has a local extremum at x_0 and f is differentiable at x_0, then $f'(x_0) = 0$.*

Proof. Assume that f is defined on $I = (a, b)$, where $a < x_0 < b$, and f (or $-f$) has a local maximum at x_0. First, we consider $f'(x_0) > 0$. Since

$$f'(x_0) = \lim_{x \to x_0} \frac{f(x) - f(x_0)}{x - x_0},$$

there exists $\delta > 0$ such that $a < x_0 - \delta < x_0 + \delta < b$ and

$$0 < |x - x_0| < \delta \implies \frac{f(x) - f(x_0)}{x - x_0} > 0. \tag{7.3.1}$$

If we select x so that $x_0 < x < x_0 + \delta$, then (7.3.1) shows that $f(x) > f(x_0)$, which contradicts the assumption that f has a local maximum at x_0.

Similarly, if $f'(x_0) < 0$, there exists $\delta > 0$ such that

$$0 < |x - x_0| < \delta \implies \frac{f(x) - f(x_0)}{x - x_0} < 0. \tag{7.3.2}$$

If we select x so that $x_0 - \delta < x < x_0$, then (7.3.2) shows that $f(x) > f(x_0)$, which is a contradiction. Thus, we must have $f'(x_0) = 0$. This completes the proof.

If $f : [a, b] \to \mathbb{R}$ is a continuous function, then f is a bounded function. Moreover, f has a maximum value and a minimum value on $[a, b]$. Thus, by using Theorem 7.3.1, we have the following.

Theorem 7.3.2. (Rolle's Theorem) *If f is a continuous function on $[a, b]$ and f is a differentiable function on (a, b) with $f(a) = f(b)$, then there exists at least one $x \in (a, b)$ such that $f'(x) = 0$.*

Proof. It is known that there exist $x_0, y_0 \in [a, b]$ such that

$$f(x_0) \le f(x) \le f(y_0) \text{ for every } x \in [a, b].$$

If x_0 and y_0 are both endpoints of $[a, b]$, then f is a constant function since $f(a) = f(b)$ and $f'(x) = 0$ for every $x \in (a, b)$. Otherwise, f has either a maximum value or a minimum value at a point $x \in (a, b)$ and so, in which case, $f'(x) = 0$ by Theorem 7.3.1. This completes the proof.

Note the following:

(1) The point $x_0 \in (a, b)$ guaranteed by Rolle's Theorem is not necessarily *unique*. For example, consider a function

$$f(x) = 3x^4 - 6x^2 + 1.$$

Then, $f(-2) = f(2) = 25$ and $f'(x) = 12x^3 - 12x$ and so

$$f'(-1) = f'(0) = f'(1);$$

(2) The assumption $f(a) = f(b)$ implies that the chord joining two points $(a, f(a))$ and $(b, f(b))$ on the graph of f is horizontal. Then, Rolle's Theorem implies that if this is the case, then the tangent line to the graph of f is horizontal at some point $(c, f(c))$ with $c \in (a, b)$;

(3) If we rotate the graph of f in Rolle's Theorem, then the assumption $f(a) = f(b)$ will no longer be satisfied and so the chord joining two points $(a, f(a))$ and $(b, f(b))$ will not be horizontal. But, the new graph of f will have the property that the tangent line will be parallel to the chord at least once between the endpoints $(a, f(a))$ and $(b, f(b))$.

Example 7.3.1. Let $f : \left[-\frac{3}{2}, \frac{3}{2} \right] \to \mathbb{R}$ be a function defined by

$$f(x) = 4x^3 - 9x \text{ for every } x \in \left[-\frac{3}{2}, \frac{3}{2} \right].$$

By Rolle's Theorem, find some points $x_0 \in \left[-\frac{3}{2}, \frac{3}{2} \right]$ for which $f'(x_0) = 0$.

Solution. It is easy to see that the function $f(x) = 4x^3 - 9x$ satisfies all the conditions of Rolle's Theorem. To find suitable points $x_0 \left[-\frac{3}{2}, \frac{3}{2} \right]$, set $f'(x) = 0$ and get

$$f'(x) = 12x^2 - 9,$$

which gives

$$x_0 = -\frac{1}{2}\sqrt{3}, \ \frac{1}{2}\sqrt{3} \in \left[-\frac{3}{2}, \frac{3}{2}\right].$$

Now, we apply Rolle's Theorem to prove one of the most important theorems in real analysis, which is well known as the Mean Value Theorem, which tells us that a differentiable function on $[a, b]$ must somewhere have its derivative equal to the slope of the line connecting $(a, f(a))$ to $(b, f(b))$, that is, $\frac{f(b)-f(a)}{b-a}$. The Mean Value Theorem is used to prove many theorems of both differential and integral calculus as well as numerical analysis and others.

Theorem 7.3.2. (Lagrange's Mean Value Theorem) *If f is a continuous function on $[a, b]$ and f is a differentiable function on (a, b), then there exists at least one $x_0 \in (a, b)$ such that*

$$f'(x_0) = \frac{f(b) - f(a)}{b - a}.$$

Proof. Consider the function

$$F(x) = f(x) - f(a) - \frac{f(b) - f(a)}{x - a} \quad \text{for every } x \in [a, b],$$

where g is simply the difference between the function f and the affine function

$$x \mapsto \frac{f(b) - f(a)}{x - a} + f(a) \quad \text{for every } x \in [a, b],$$

whose graph is the line segment joining the points $(a, f(a))$ and $(b, f(b))$.

Now, we show that F satisfies all the conditions of Rolle's Theorem. In fact, it is very easy to see that F is continuous on $[a, b]$, differentiable on (a, b) and $F(a) = F(b)$. So, by Rolle's Theorem, there exists $x_0 \in (a, b)$ such that

$$F'(x_0) = 0.$$

But we have

$$F'(x) = f'(x) - \frac{f(b) - f(a)}{b - a}$$

and so

$$0 = F'(x_0) = f'(x_0) - \frac{f(b) - f(a)}{b - a},$$

which implies

$$f'(x_0) = \frac{f(b) - f(a)}{b - a}.$$

This completes the proof.

Note that Rolle's Theorem is a special case of the Mean Value Theorem, where $f(b) = f(a)$.

Example 7.3.1. Let $f : [1,3] \to \mathbb{R}$ be a function defined by

$$f(x) = x^3 - 5x^2 - 3x \quad \text{for every} \quad x \in [1,3].$$

Show that all the conditions of the Mean Value Theorem are satisfied, and find every point $x_0 \in (1,3)$ such that

$$f'(x_0) = \frac{f(3) - f(1)}{3 - 1}.$$

Solution. It is easy to show that f is continuous on $[1,3]$ and differentiable on $(1,3)$ and so all the conditions of the Mean Value Theorem are satisfied.

Next, for the function $f(x) = x^3 - 5x^2 - 3x$, we have the following:

$$f'(x) = 3x^2 - 10x - 3, \quad f(1) = -7, \quad f(3) = -27$$

and so

$$\frac{f(3) - f(1)}{3 - 1} = \frac{-27 - (-7)}{2} = -10.$$

Set $f'(x_0) = -10$ to obtain the following:

$$3x_0^2 - 10x_0 - 3 = -10, \quad (3x_0 - 7)(x_0 - 1) = 0,$$

which gives

$$x_0 = \frac{7}{3}, \quad x_0 = 1.$$

But, since $1 \neq (1,3)$, the only one value of x_0 is $\frac{7}{3}$.

By using Lagrange's Mean Value Theorem, we have the following.

Corollary 7.3.3. *If a function $f : (a,b) \to \mathbb{R}$ is differentiable on (a,b) such that $f'(x) = 0$ for all $x \in (a,b)$, then f is a constant function on (a,b), that is, $f(x) = C$, where C is a constant.*

Proof. Suppose that f is not constant on (a,b). Then, there exist $x_1, x_2 \in (a,b)$ with $a < x_1 < x_2 < b$ such that $f(x_1) \neq f(x_2)$. By Lagrange's Mean Value Theorem, there exists $x_0 \in (x_1, x_2)$ such that

$$f'(x_0) = \frac{f(x_2) - f(x_1)}{x_2 - x_1} \neq 0,$$

which is a contradiction. Thus, the conclusion follows. This completes the proof.

Corollary 7.3.4. *If two functions $f, g : (a,b) \to \mathbb{R}$ are differentiable on (a,b) such that $f'(x) = g'(x)$ for every $x \in (a,b)$, then there exists a constant C such that*

$$f(x) = g(x) + C \quad \text{for every} \quad x \in (a,b).$$

Proof. Let $h(x) = f(x) - g(x)$ for every $x \in (a, b)$. Then, h is differentiable on (a, b) and

$$h^{'}(x) = f^{'}(x) - g^{'}(x) = 0 \ \text{ for every } \ x \in (a, b).$$

So, by Corollary 7.3.3, we have $h(x) = C$, where C is a constant, which implies that

$$f(x) = g(x) + C \ \text{ for every } \ x \in (a, b).$$

This completes the proof.

Corollary 7.3.5. *If $f : (a, b) \to \mathbb{R}$ is differentiable on an interval (a, b), then we have the following:*

(1) *f is strictly increasing on (a, b) if $f^{'}(x) > 0$ for every $x \in (a, b)$;*

(2) *f is strictly decreasing on (a, b) if $f^{'}(x) < 0$ for every $x \in (a, b)$;*

(3) *f is increasing on (a, b) if $f^{'}(x) \geq 0$ for every $x \in (a, b)$;*

(4) *f is decreasing on (a, b) if $f^{'}(x) \leq 0$ for every $x \in (a, b)$.*

Proof. (1) Let $x_1, x_2 \in (a, b)$ with $x_1 < x_2$. Then, by Lagrange's Mean Value Theorem, there exists $x_0 \in (x_1, x_2)$ such that

$$\frac{f(x_2) - f(x_1)}{x_2 - x_1} = f^{'}(x_0) > 0.$$

Since $x_2 - x_l > 0$, it follows that $f(x_2) - f(x_1) > 0$ or $f(x_2) > f(x_1)$. This implies that f is strictly increasing on (a, b).

(2) Let $x_1, x_2 \in (a, b)$ with $x_1 < x_2$. Then, by Lagrange's Mean Value Theorem, there exists $x_0 \in (x_1, x_2)$ such that

$$\frac{f(x_2) - f(x_1)}{x_2 - x_1} = f^{'}(x_0) < 0.$$

Since $x_2 - x_l > 0$, it follows that $f(x_2) - f(x_1) < 0$ or $f(x_2) < f(x_1)$. This implies that f is strictly decreasing on (a, b).

Similarly, we have conclusions (3) and (4). This completes the proof.

Note that the derivative $f^{'}$ of a differentiable function f need not be continuous. But, the derivative f' has the Intermediate Value Theorem as follows.

Theorem 7.3.6. (The Intermediate Value Theorem for Derivatives) *Let f be a differentiable function on (a, b). If $x_1, x_2 \in (a, b)$ with $x_1 < x_2$ and c is between $f^{'}(x_1)$ and $f^{'}(x_2)$, then there exists at least one $x_0 \in (x_1, x_2)$ such that $f'(x_0) = c$.*

Proof. Since c is between $f^{'}(x_1)$ and $f^{'}(x_2)$, we assume that

$$f^{'}(x_1) < c < f^{'}(x_2).$$

Let $g(x) = f(x) - cx$ for every $x \in (x_1, x_2)$. Then, g is differentiable on (x_1, x_2) and so g is continuous on $[x_1, x_2]$. So, g attains its minimum on $[x_1, x_2]$, that is, for some point $x_0 \in [x_1, x_2]$, we have $g(x_0) \le g(x)$ for every $x \in [x_1, x_2]$.

On the other hand, since $g'(x_1) = f'(x_1) - c$, we have $g'(x_1) < 0$ and also $g'(x_2) > 0$. Thus, it follows the definition of the derivative that, for every $\delta > 0$ small enough, $g(x) < g(x_1)$ for every $x \in (x_1, x_1 + \delta)$, and $g(x) < g(x_2)$ for every $x \in (x_2 - \delta, x_2)$. Therefore, the minimum value of g on $[x_1, x_2]$ occurs at a point $x_0 \in (x_1, x_2)$. Therefore, by Fermat's Theorem, $g'(x_0) = 0$, that is, $0 = g'(x_0) = f'(x_0) - c$, which implies that $f'(x_0) = c$. This completes the proof.

Example 7.3.2. Consider a function $g : \mathbb{R} \to \mathbb{R}$ defined by

$$g(x) = \begin{cases} x^2 \sin \frac{1}{x}, & \text{if } x \ne 0, \\ 0, & \text{if } x = 0. \end{cases}$$

Then, g is differentiable on $\mathbb{R}$, and, in fact, g' is continuous except at $x = 0$. But, despite the discontinuity of g' at $x = 0$, the function g' has the Intermediate Value Property.

Theorem 7.3.7. (Cauchy's Mean Value Theorem) *If $f, g : [a, b] \to \mathbb{R}$ are continuous on $[a, b]$ and are differentiable on (a, b) and $g'(x) \ne 0$ for every $x \in (a, b)$, then there exists at least one $x_0 \in (a, b)$ such that*

$$\frac{f'(x_0)}{g'(x_0)} = \frac{f(b) - f(a)}{g(b) - g(a)}.$$

Proof. Consider the function $F :: [a, b] \to \mathbb{R}$ defined by

$$F(x) = f(x)[g(b) - g(a)] - g(x)[f(b) - f(a)] \quad \text{for every } x \in [a, b]. \quad (7.3.3)$$

Now, by using Rolle's Theorem, we show that $F'(x_0) = 0$ for some $x_0 \in (a, b)$. In fact, it follows from (7.3.3) that F is continuous on $[a, b]$, F is differentiable on (a, b), and $F(a) = F(b)$. Thus, by Rolle's Theorem, there exists a point $x_0 \in (a, b)$ such that $F'(x_0) = 0$, which implies that

$$0 = F'(x_0) = f'(x_0)[g(b) - g(a)] - g'(x_0)[f(b) - f(a)].$$

Therefore, we have

$$\frac{f'(x_0)}{g'(x_0)} = \frac{f(b) - f(a)}{g(b) - g(a)}.$$

This completes the proof.

Note that, if $g(x) = x$ for every $x \in [a, b]$ in Cauchy's Mean Value Theorem, then we can get Lagrange's Mean Value Theorem.

Example 7.3.3. Consider the following two functions $f, g : [0, 1] \to \mathbb{R}$ defined by

$$f(x) = 3x^2 + 3x - 1, \quad g(x) = x^3 - 4x + 2 \text{ for every } x \in [0, 1].$$

Then, f and g are continuous on $[0, 1]$, differentiable on $(0, 1)$, and $g'(x) \neq 0$ for every $x \in (0, 1)$. Also, we have

$$f'(x) = 6x + 3, \quad g'(x) = 3x^2 - 4$$

and

$$f(1) = 5, \quad g(1) = -1, \quad f(0) = -1, \quad g(0) = 2.$$

Thus, by Cauchy's Mean Value Theorem, there exists $x_0 \in (0, 1)$ such that

$$\frac{f(1) - f(0)}{g(1) - g(0)} = \frac{6x_0 + 3}{3x_0^2 - 4},$$

which implies that

$$\frac{5 - (-1)}{-1 - 2} = \frac{6x_0 + 3}{3x_0^2 - 4}, \quad 6x_0^2 + 6x_0 - 5 = 0.$$

Therefore, we have

$$x_0 = \frac{-6 \pm \sqrt{36 + 120}}{12} = \frac{-6 \pm 2\sqrt{39}}{12} = \frac{-3 \pm \sqrt{39}}{6}.$$

Therefore, only one number is $x_0 = \frac{1}{6}(-3 + \sqrt{39}) \in (0, 1)$.

7.4 The L'Hospital Rule

For example, if $\lim_{x \to x_0} f(x) = L$ and $\lim_{x \to x_0} g(x) = M$, then we can consider the following cases:

(1) If $\lim_{x \to x_0} g(x) = M \neq 0$, then

$$\lim_{x \to x_0} \frac{f(x)}{g(x)} = \frac{L}{M};$$

(2) If $\lim_{x \to x_0} g(x) = M = 0$, then no conclusion was deduced;

(3) If $\lim_{x \to x_0} f(x) = L \neq 0$ and $\lim_{x \to x_0} g(x) = M = 0$, then

$$\lim_{x \to x_0} \frac{f(x)}{g(x)} = \infty;$$

(4) If $\lim_{x \to x_0} f(x) = L = 0$ and $\lim_{x \to x_0} g(x) = M = 0$, then the limit $\lim_{x \to x_0} \frac{f(x)}{g(x)}$ has not been covered;

(5) If $\lim_{x \to x_0} f(x) = L = \infty$ and $\lim_{x \to x_0} g(x) = M = \infty$, then the limit $\lim_{x \to x_0} \frac{f(x)}{g(x)}$ has not been covered.

In cases (4) and (5), the limit of the quotient $\frac{f}{g}$ is said to be *indeterminate*.

Definition 7.4.1. Let I be an interval or an unbounded interval, and let $x_0 \in I$ or $x_0 = \pm\infty$ if I is unbounded. If f and g are defined on I, then the *indeterminate forms*

$$\frac{0}{0}, \quad \frac{\infty}{\infty}, \quad 0 \cdot \infty, \quad \infty - \infty, \quad 0^0, \quad 1^\infty, \quad \infty^0$$

are defined as follows:

(1) The limit $\lim_{x \to x_0} \frac{f(x)}{g(x)}$ is said to be the *indeterminate form* $\frac{0}{0}$ if

$$\lim_{x \to x_0} f(x) = 0 = \lim_{x \to x_0} g(x);$$

(2) The limit $\lim_{x \to x_0} \frac{f(x)}{g(x)}$ is said to be the *indeterminate form* $\frac{\infty}{\infty}$ if

$$|\lim_{x \to x_0} f(x)| = \infty = |\lim_{x \to x_0} g(x)|;$$

(3) The limit $\lim_{x \to x_0} f(x) \cdot g(x)$ is said to be the *indeterminate form* $0 \cdot \infty$ if

$$\lim_{x \to x_0} f(x) = 0, \quad \lim_{x \to x_0} g(x) \pm \infty$$

or

$$\lim_{x \to x_0} g(x) = 0, \quad \lim_{x \to x_0} f(x) \pm \infty;$$

(4) The limit $\lim_{x \to x_0} f(x) - g(x)$ is said to be the *indeterminate form* $\infty - \infty$ if

$$\lim_{x \to x_0} f(x) \pm \infty = \lim_{x \to x_0} g(x) = 0;$$

(5) The limit $\lim_{x \to x_0} f(x)^{g(x)}$ is said to be the *indeterminate form* 0^0 if

$$\lim_{x \to x_0} f(x) = 0 = \lim_{x \to x_0} g(x);$$

(6) The limit $\lim_{x \to x_0} f(x)^{g(x)}$ is said to be the *indeterminate form* 1^∞ if

$$\lim_{x \to x_0} f(x) = 1, \quad \lim_{x \to x_0} g(x) \pm \infty;$$

(7) The limit $\lim_{x \to x_0} f(x)^{g(x)}$ is said to be the *indeterminate form* ∞^0 if
$$\lim_{x \to x_0} f(x) = \infty, \quad \lim_{x \to x_0} g(x) = 0;$$

(8) For every $x_0 \in \mathbb{R}$, indeterminate forms (1)–(7) can also be defined for one-sided limits $x \to x_0^+$ and $x \to x_0^-$.

Theorem 7.4.1. *Let $f, g : [a, b] \to \mathbb{R}$ be two functions on $[a, b]$ with $f(a) = g(a) = 0$ and $g(x) \neq 0$ for every $x \in (a, b)$. If f and g are differentiable at a and $g'(a) \neq 0$, then*

(1) $\lim_{x \to a^+} \frac{f(x)}{g(x)}$ *exists;*

(2) $\lim_{x \to a^+} \frac{f(x)}{g(x)} = \frac{f'(a)}{g'(a)}$.

Proof. Since $f(a) = g(a) = 0$, for every $x \in (a, b)$, we have the following:
$$\frac{f(x)}{g(x)} = \frac{f(x) - f(a)}{g(x) - g(a)} = \frac{\frac{f(x) - f(a)}{x - a}}{\frac{g(x) - g(a)}{x - a}}.$$

Therefore, we have
$$\lim_{x \to a^+} \frac{f(x)}{g(x)} = \frac{\lim_{x \to a^+} \frac{f(x) - f(a)}{x - a}}{\lim_{x \to a^+} \frac{g(x) - g(a)}{x - a}} = \frac{f'(a)}{g'(a)}.$$

This completes the proof.

Note that the condition $f(a) = g(a) = 0$ is essential. For example, if $f(x) = x + 17$ and $g(x) = 2x + 3$ for every $x \in \mathbb{R}$, then we have
$$\lim_{x \to 0} \frac{f(x)}{g(x)} = \frac{17}{3},$$

but
$$\frac{f'(0)}{g'(0)} = \frac{1}{2},$$

which implies that
$$\lim_{x \to 0} \frac{f(x)}{g(x)} \neq \frac{f'(0)}{g'(0)}.$$

Theorem 7.4.1. (The L'Hospital Rule: $\frac{0}{0}$) *Let f and g be continuous functions on $[a, b]$ and differentiable on (a, b) with $f(a) = g(a) = 0$, $g(x) \neq 0$, and $g'(x) \neq 0$ for every $x \in (a, b)$. Then, we have the following:*

(1) *If $\lim_{x \to a^+} \frac{f'(x)}{g'(x)} = L$ for some $L \in \mathbb{R}$, then $\lim_{x \to a^+} \frac{f(x)}{g(x)} = L$;*

(2) *If $\lim_{x \to a^+} \frac{f'(x)}{g'(x)} = \infty$ (resp., $-\infty$), then $\lim_{x \to a^+} \frac{f(x)}{g(x)} = \infty$ (resp., $-\infty$).*

Proof. (1) Since $\lim_{x \to a^+} \frac{f'(x)}{g'(x)} = L$, it follows that, for every $\varepsilon > 0$, there exists $\delta > 0$ such that

$$a < x < a + \delta \implies \left| \frac{f'(x)}{g'(x)} - L \right| < \varepsilon.$$

Since f and g satisfy all the conditions of Cauchy's Mean Value Theorem on $[a, x]$, for every $x \in (a, a + \delta)$, there exists $c_x \in (a, x)$ such that

$$\frac{f(x)}{g(x)} = \frac{f(x) - f(a)}{g(x) - g(a)} = \frac{f'(c_x)}{g'(c_x)}.$$

Thus, for every $x \in (a, a + \delta)$, we have

$$\left| \frac{f(x)}{g(x)} - L \right| = \left| \frac{f'(c_x)}{g'(c_x)} - L \right| < \varepsilon.$$

So, we have

$$\lim_{x \to a^+} \frac{f(x)}{g(x)} = L.$$

(2) Let $\lim_{x \to a^+} \frac{f'(x)}{g'(x)} = \infty$. Then, for every $K > 0$, there exists $\delta > 0$ such that

$$a < x < a + \delta \implies \frac{f'(x)}{g'(x)} > K.$$

On the other hand, as in (1), for every $x \in (a, a + \delta)$, there exists $c_x \in (a, x)$ such that

$$\frac{f(x) - f(a)}{g(x) - g(a)} = \frac{f(x)}{g(x)} = \frac{f'(c_x)}{g'(c_x)}.$$

Therefore, for every $x \in (a, a + \delta)$, since $\frac{f'(x)}{g'(x)} > K$, we have

$$\frac{f(x)}{g(x)} > K,$$

which implies that

$$\lim_{x \to a^+} \frac{f(x)}{g(x)} = \infty.$$

This completes the proof.

Note that Theorem 7.4.1 is established for right-hand limits, that is, as $x \to x_0^+$, but it is also clear for left-hand limits, that is, as $x \to x_0^-$. Thus, combining the results for right-hand and left-hand limits, the corresponding results for $x \to x_0$ are established.

From Theorem 7.4.1, we have the following:

Corollary 7.4.2. *Let f and g be differentiable on $[a, \infty)$ with $g'(x) \neq 0$ and*

$$\lim_{x \to \infty} f(x) = \lim_{x \to \infty} g(x) = 0.$$

Then, if $\lim_{x \to \infty} \frac{f'(x)}{g'(x)} = L$ for some $L \in \mathbb{R}$, then $\lim_{x \to \infty} \frac{f(x)}{g(x)} = L$.

Proof. Let $a > 0$ and $b = \infty$. Then, define the functions $F, G : [0, \frac{1}{a}] \to \mathbb{R}$ by

$$F(t) = \begin{cases} f(\frac{1}{t}), & \text{if } 0 < t \leq \frac{1}{a}, \\ 0, & \text{if } t = 0, \end{cases}$$

and

$$G(t) = \begin{cases} g(\frac{1}{t}), & \text{if } 0 < t \leq \frac{1}{a}, \\ 0, & \text{if } t = 0. \end{cases}$$

Then, it follows that F and G are differentiable on $(0, \frac{1}{a})$, $G'(t) \neq 0$, and for every $t \in (0, \frac{1}{a})$,

$$F'(t) = \left(-\frac{1}{t^2}\right) f'\left(\frac{1}{t}\right), \quad G'(t) = \left(-\frac{1}{t^2}\right) g'\left(\frac{1}{t}\right),$$

$$\lim_{t \to 0^+} F(t) = \lim_{x \to \infty} f(x), \quad \lim_{t \to 0^+} G(t) = \lim_{x \to \infty} g(x)$$

and

$$\lim_{x \to 0^+} \frac{F'(x)}{G'(x)} = \lim_{x \to 0^+} \frac{f'(\frac{1}{x})}{g'(\frac{1}{x})} = \lim_{x \to \infty} \frac{f'(x)}{g'(x)} = L.$$

Therefore, by Theorem 7.4.1, we have

$$\lim_{x \to \infty} \frac{f'(x)}{g'(x)} = \lim_{x \to 0^+} \frac{F'(x)}{G'(x)} = L = \lim_{x \to 0^+} \frac{F(x)}{G(x)} = \lim_{x \to \infty} \frac{f(x)}{g(x)}.$$

This completes the proof.

Note that, in the case $x \to -\infty$, the conclusion of Theorem 7.4.2 follows.

Example 7.4.1. Evaluate the following limits:

(1) $\lim_{x \to 0} \frac{1 - \cos x}{x^2}$;

(2) $\lim_{x \to \infty} x\left(a^{\frac{1}{x}} - 1\right)$, where $a > 0$;

(3) $\lim_{x \to \infty} \frac{x + \sin x \cos x}{x \sin x + \cos x}$.

Solution. (1) We have

$$\lim_{x \to 0} \frac{1 - \cos x}{x^2} = \lim_{x \to 0} \frac{\sin x}{2x}.$$

The quotient in the second limit is again indeterminate in the form $\frac{0}{0}$. So, we have to apply L'Hospital Rule in the second limit again. Thus, we have

$$\lim_{x \to 0} \frac{1 - \cos x}{x^2} = \lim_{x \to 0} \frac{\sin x}{2x} = \lim_{x \to 0} \frac{\cos x}{2} = \frac{1}{2}.$$

(2) We have

$$\begin{aligned}
\lim_{x \to \infty} x\left(a^{\frac{1}{x}} - 1\right) &= \lim_{x \to \infty} \frac{a^{\frac{1}{x}} - 1}{\frac{1}{x}} \\
&= \lim_{x \to \infty} \frac{a^{\frac{1}{x}}\left(-\frac{1}{x^2}\right) \ln a}{-\frac{1}{x^2}} \\
&= \lim_{x \to \infty} a^{\frac{1}{x}} \ln a \\
&= \ln a.
\end{aligned}$$

(3) Let $f(x) = x + \sin x \cos x$ and $g(x) = x \sin x + \cos x$. If $x = 2n\pi + \frac{\pi}{2}$, then $\frac{f(x)}{g(x)} = 1$, and if $x = 2n\pi - \frac{\pi}{2}$, then $\frac{f(x)}{g(x)} = -1$, and so $\lim_{x \to \infty} \frac{f(x)}{g(x)}$ does not exist. Further, $g'(x) = x \cos x$, and if $x = \frac{(2n-1)\pi}{2}$, then $g'(x) = 0$, and so we cannot apply L'Hospital Rule.

Theorem 7.4.3. (The L'Hospital Rule: $\frac{\infty}{\infty}$) *Let f and g be differentiable on (a, b) with $g(x) \neq 0$, $g'(x) \neq 0$ for every $x \in (a, b)$ and*

$$\lim_{x \to a^+} f(x) = \infty = \lim_{x \to a^+} g(x).$$

Then, we have the following:

(1) *If $\lim_{x \to a^+} \frac{f'(x)}{g'(x)} = L$ for some $L \in \mathbb{R}$, then $\lim_{x \to a^+} \frac{f(x)}{g(x)} = L$;*

(2) *If $\lim_{x \to a^+} \frac{f'(x)}{g'(x)} = \infty$ (resp., $-\infty$), then $\lim_{x \to a^+} \frac{f(x)}{g(x)} = \infty$ (resp., $-\infty$).*

Proof. (1) Let $\lim_{x \to a^+} \frac{f'(x)}{g'(x)} = L$ for some $L \in \mathbb{R}$. Then, for every $\varepsilon > 0$ with $0 < \varepsilon < 1$, there exists $\delta \in (0, b - a)$, that is, $0 < \delta < b - a$, such that

$$a < x < a + \delta \implies \left|\frac{f'(x)}{g'(x)} - L\right| < \frac{\varepsilon}{2} \tag{7.4.1}$$

and so

$$a < x < a + \delta \implies \left|\frac{f'(x)}{g'(x)}\right| < |L| + \frac{\varepsilon}{2} < |L| + \frac{1}{2}. \tag{7.4.2}$$

On the other hand, since $\lim_{x \to a^+} f(x) = \infty$, there exists $\delta_1 > 0$ with $\delta_1 < \delta$ such that

$$a < x < a + \delta_1 \implies f(x) > |f(a + \delta)|.$$

Now, for every $x \in (a, a + \delta_1)$, if we apply Cauchy's Mean Value Theorem to the functions f and g on $[a, a + \delta]$, then there exists $c_x \in (x, a + \delta)$ such that

$$\frac{f(x) - f(a + \delta)}{g(x) - g(a + \delta)} = \frac{f(x)}{g(x)} \cdot \frac{1 - \frac{f(a+\delta)}{f(x)}}{1 - \frac{g(a+\delta)}{g(x)}} = \frac{f'(c_x)}{g'(c_x)}. \tag{7.4.3}$$

Thus, we have

$$\frac{f(x)}{g(x)} = \frac{f'(c_x)}{g'(c_x)} \cdot \frac{1 - \frac{g(a+\delta)}{g(x)}}{1 - \frac{f(a+\delta)}{f(x)}}. \tag{7.4.4}$$

Since $\lim_{x \to a^+} f(x) = \infty = \lim_{x \to a^+} g(x)$, we have

$$\lim_{x \to a^+} \frac{1 - \frac{g(a+\delta)}{g(x)}}{1 - \frac{f(a+\delta)}{f(x)}} = 1.$$

Hence, it follows that there exists $\delta_2 > 0$ with $\delta_2 < \delta_1$ such that

$$a < x < a + \delta \implies \left| \frac{1 - \frac{g(a+\delta)}{g(x)}}{1 - \frac{f(a+\delta)}{f(x)}} - 1 \right| < \frac{\varepsilon}{2(|L| + \frac{1}{2})}. \tag{7.4.5}$$

Therefore, it follows from (7.4.1)–(7.4.5) that, for every $x \in (a, a + \delta_2)$,

$$\left| \frac{f(x)}{g(x)} - L \right| \leq \left| \frac{f'(c_x)}{g'(c_x)} \right| \cdot \left| \frac{1 - \frac{g(a+\delta)}{g(x)}}{1 - \frac{f(a+\delta)}{f(x)}} - 1 \right| + \left| \frac{f'(c_x)}{g'(c_x)} - L \right|$$

$$< \left(|L| + \frac{1}{2} \right) \cdot \frac{\varepsilon}{2(|L| + \frac{1}{2})} + \frac{\varepsilon}{2}$$

$$= \varepsilon$$

and so

$$\lim_{x \to a^+} \frac{f(x)}{g(x)} = \lim_{x \to a^+} \frac{f'(x)}{g'(x)} = L.$$

(2) We omit the proof here. This completes the proof.

Example 7.4.2. Evaluate the limit $\lim_{x \to 0^+} \frac{\ln x}{\frac{1}{x}}$.

Solution. By Theorem 7.4.3, we have

$$\lim_{x \to 0^+} \frac{\ln x}{\frac{1}{x}} = \lim_{x \to 0^+} \frac{\frac{1}{x}}{-\frac{1}{x^2}} = \lim_{x \to 0^+} (-x) = 0.$$

Theorem 7.4.4. *Let f and g be differentiable on (a, ∞) with $g(x) \neq 0$, $g'(x) \neq 0$ for every $x \in (a, \infty)$ and*

$$\lim_{x \to \infty} f(x) = \infty = \lim_{x \to \infty} g(x).$$

If $\lim_{x \to \infty} \frac{f'(x)}{g'(x)} = L$ for some $L \in \mathbb{R}$, then $\lim_{x \to \infty} \frac{f(x)}{g(x)} = L$.

Example 7.4.3. Evaluate the limit $\lim_{x \to \infty} \frac{\ln(2 + e^x)}{3x}$.

Solution. Since

$$\lim_{x \to \infty} \ln(2 + e^x) = \infty, \quad \lim_{x \to \infty} 3x = \infty,$$

by Theorem 7.4.4, we have

$$\lim_{x \to \infty} \frac{\ln(2 + e^x)}{3x} = \lim_{x \to \infty} \frac{\frac{1}{2+e^x} \cdot e^x}{3} = \lim_{x \to \infty} \frac{e^x}{6 + 3e^x}. \qquad (7.4.6)$$

Since

$$\lim_{x \to \infty} e^x = \infty, \quad \lim_{x \to \infty} (6 + 3e^x) = \infty,$$

we apply Theorem 7.4.4 to the third limit in (7.4.6) again as follows:

$$\lim_{x \to \infty} \frac{e^x}{6 + 3e^x} = \lim_{x \to \infty} \frac{e^x}{3e^x} = \lim_{x \to \infty} \frac{1}{3} = \frac{1}{3}.$$

Therefore, we have

$$\lim_{x \to \infty} \frac{\ln(2 + e^x)}{3x} = \frac{1}{3}.$$

Now, we consider the following *indeterminate forms* by examples:

$$0 \cdot \infty, \quad \infty - \infty, \quad 0^0, \quad 1^\infty, \quad \infty^0.$$

Example 7.4.4. Evaluate the following limits:

(1) $\lim_{x \to 0} \left(\frac{1}{x} - \frac{1}{\sin x} \right)$;

(2) $\lim_{x \to 0+} x \ln x$;

(3) $\lim_{x \to 0+} x^x$;

(4) $\lim_{x \to \infty} \left(1 + \frac{1}{x} \right)^x$;

(5) $\lim_{x \to 0+} \left(1 + \frac{1}{x} \right)^x$.

Solution. (1) The limit $\lim_{x\to 0}\left(\frac{1}{x}-\frac{1}{\sin x}\right)$ has the indeterminate form $\infty-\infty$. Now, we have

$$\lim_{x\to 0}\left(\frac{1}{x}-\frac{1}{\sin x}\right) = \lim_{x\to 0}\frac{\sin x - x}{x\sin x}$$
$$= \lim_{x\to 0}\frac{\cos x - 1}{\sin x + x\cos x}$$
$$= \lim_{x\to 0}\frac{-\sin x}{\cos x + \cos x - x\sin x}$$
$$= \frac{0}{2} = 0.$$

(2) The limit $\lim_{x\to 0+} x\ln x$ has the indeterminate form $0\cdot(-\infty)$. Now, we have

$$\lim_{x\to 0+} x\ln x = \lim_{x\to 0+}\frac{\ln x}{\frac{1}{x}} = \lim_{x\to 0+}\frac{\frac{1}{x}}{-\frac{1}{x^2}} = \lim_{x\to 0+}(-x) = 0.$$

(3) The limit $\lim_{x\to 0+} x^x$ has the indeterminate form 0^0. Now, we have

$$x^x = e^{x\ln x}$$

and so, since the function $f(x) = e^x$ is continuous at $x = 0$, by (2), we have

$$\lim_{x\to 0+} x^x = \lim_{x\to 0+} e^{x\ln x} = e^{\lim_{x\to 0+} x\ln x} = e^0 = 1.$$

(4) The limit $\lim_{x\to\infty}\left(1+\frac{1}{x}\right)^x$ has the indeterminate form 1^∞. Now, we have

$$\left(1+\frac{1}{x}\right)^x = e^{x\ln(1+\frac{1}{x})} \tag{7.4.7}$$

and

$$\lim_{x\to\infty} x\ln\left(1+\frac{1}{x}\right) = \lim_{x\to\infty}\frac{\ln(1+\frac{1}{x})}{\frac{1}{x}}$$
$$= \lim_{x\to\infty}\frac{(1+\frac{1}{x})^{-1})(-x^{-2})}{-x^{-2}}$$
$$= \lim_{x\to\infty}\frac{1}{1+\frac{1}{x}}$$
$$= 1.$$

Therefore, since $f(x) = e^x$ is continuous at $x = 1$, we have

$$\lim_{x\to\infty}\left(1+\frac{1}{x}\right)^x = e.$$

(5) The limit $\lim_{x\to 0+}\left(1+\frac{1}{x}\right)^x$ has the indeterminate form ∞^0. Now, from (7.4.7), we have

$$\lim_{x \to 0+} x \ln\left(1 + \frac{1}{x}\right) = \lim_{x \to 0+} \frac{\ln(1 + \frac{1}{x})}{\frac{1}{x}}$$

$$= \lim_{x \to 0+} \frac{1}{1 + \frac{1}{x}}$$

$$= 0.$$

7.5 Taylor's Theorem

First, we introduce the nth derivative of a given function.

The *first derivative* of the function is defined as follows:

$$f'(x) = \lim_{\Delta x \to 0} \frac{f(x + \Delta x) - f(x)}{\Delta x}.$$

Also, f' is a function. So, if the first derivative of the function f exists, then we can define the *second derivative* of the first derivative f' as follows:

$$f''(x) = \lim_{\Delta x \to 0} \frac{f'(x + \Delta x) - f'(x)}{\Delta x}.$$

Also, f'' is a function. So, if the second derivative of the function f' exists, then we can define the *third derivative* of the second derivative f'' as follows:

$$f'''(x) = \lim_{\Delta x \to 0} \frac{f''(x + \Delta x) - f''(x)}{\Delta x}.$$

Inductively, if the $(n - 1)$th derivative of the function f^{n-2} exists, then we can define the *nth derivative* of the function f as follows:

$$f^{(n)}(x) = \lim_{\Delta x \to 0} \frac{f^{(n-1)}(x + \Delta x) - f^{(n-1)}(x)}{\Delta x}.$$

Definition 7.5.1. (1) If the nth derivative f^n of a function $f : (a, b) \to \mathbb{R}$ exists and is continuous on (a, b), then the function f is called the *function of class C^n*, where $C^n(a, b)$ (simply, C^n) denotes the set of all functions of class C^n;

(2) For every $n \geq 1$, if f is the function of class C^n, the f is called the *function of class C^∞*, where $C^\infty(a, b)$ (simply, C^∞) denotes the set of all functions of class C^∞.

Example 7.5.1. Let a function $f : \mathbb{R} \to \mathbb{R}$ be a polynomial of degree n, that is,

$$f(x) = a_n x^n + a_{n-1} x^{n-1} + \cdots + a_1 x + a_0,$$

where $a_0, a_1, \ldots, a_{n-1}, a_n$ are coefficients. Then, f is differentiable on $\mathbb{R}$ and the function of class C^n.

Let a function $f : \mathbb{R} \to \mathbb{R}$ be defined by

$$f(x) = a_0 + a_1(x - c) + a_2(x - c)^2 + \cdots + a_n(x - c)^n + \cdots, \qquad (7.5.1)$$

where $a_0, a_1, \ldots, a_{n-1}, a_n$ are coefficients. Then, we have the following:

$$f(c) = a_0, \; f'(c) = a_1, \; f''(c) = 2a_2, \; \ldots, \; f^{(n)}(c) = n!a_n, \; \ldots$$

and so

$$a_0 = f(c), \; a_1 = f'(c), \; a_2 = \frac{f''(c)}{2!}, \; \ldots, \; a_n = \frac{f^{(n)}(c)}{n!}, \; \ldots.$$

Thus, it follows from (7.5.1) that

$$f(x) = f(c) + \frac{f'(c)}{1!}(x - c) + \frac{f''(c)}{2!}(x - c)^2 + \cdots + \frac{f^{(n)}(c)}{n!}(x - c)^n + \cdots.$$

Definition 7.5.2. Let $f : [a, b] \to \mathbb{R}$ be a function, $f \in C^n[a, b]$ and $c \in (a, b)$. Then

$$P_n(x) = f(c) + \frac{f'(c)}{1!}(x - c) + \frac{f''(c)}{2!}(x - c)^2 + \cdots + \frac{f^{(n)}(c)}{n!}(x - c)^n$$

is called the *Taylor polynomial of degree n.*

Now, we prove Taylor's Theorem as follows:

Theorem 7.5.1. (Taylor's Theorem) *Let $f \in C^{n+1}(a, b)$ and $c, x \in (a, b)$ with $c < x$. Then, there exists $d \in (c, x)$ such that*

$$f(x) = P_n(x) + \frac{(x - c)^{n+1}}{(n + 1)!} f^{(n+1)}(d).$$

Proof. For every $t \in [c, x]$, define a function $F : [c, x] \to \mathbb{R}$ by

$$F(t) = f(x) - \left(f(t) + \frac{f'(t)}{1!}(x - t) + \frac{f''(t)}{2!}(x - t)^2 \right.$$
$$\left. + \cdots + \frac{f^{(n)}(t)}{n!}(x - t)^n \right). \qquad (7.5.2)$$

Then, we have the following:

(1) F is continuous on $[c, x]$;

(2) F is differentiable on (c, x);

(3) Further, we have

$$F'(t) = -\frac{f^{(n+1)}(t)}{n!}(x-t)^n. \tag{7.5.3}$$

Now, we define a new function $G : [c, x] \to \mathbb{R}$ by

$$G(t) = F(t) - \left(\frac{x-t}{x-c}\right)^{n+1} F(c) \quad \text{for every } t \in [c, x].$$

Then, $G(c) = G(x) = 0$, G is continuous on $[c, x]$, and G is differentiable on (c, x). Thus, by Rolle's Theorem, there exists $d \in (c, x)$ such that

$$G'(d) = 0.$$

So, we have

$$G'(d) = F'(d) + \frac{(n+1)(x-d)^n}{(x-c)^{n+1}} F(c),$$

where

$$F(c) = -\frac{1}{n} \frac{(x-c)^{n+1}}{(x-d)^n} F'(d).$$

Thus, from (7.5.3), we can get $F'(d)$ and

$$F(c) = \frac{1}{n+1} \frac{(x-c)^{n+1}}{(x-d)^n} \frac{(x-d)^n}{n!} f^{(n+1)}(d) = \frac{f^{(n+1)}(d)}{(n+1)!}(x-c)^{n+1}. \tag{7.5.4}$$

Therefore, it follows from (7.5.2) and (7.5.4) that

$$f(x) = P_n(x) + \frac{(x-c)^{n+1}}{(n+1)!} f^{(n+1)}(d) = P_n(x) + \frac{f^{(n+1)}(d)}{(n+1)!}(x-c)^{n+1}.$$

This completes the proof.

Note that if $n = 0$ in Taylor's Theorem, then we have Lagrange's Mean Value Theorem.

Definition 7.5.3. For every $f \in C^{n+1}(a, b)$,

$$f(x) - P_n(x) = \frac{f^{(n+1)}(d)}{(n+1)!}(x-c)^{n+1} \quad (c < d < x)$$

is called the *Lagrange form* of the remainder, which is denoted as follows:

$$R_n(x) = \frac{f^{(n+1)}(d)}{(n+1)!}(x-c)^{n+1}.$$

Theorem 7.5.2. *Let $f \in C^\infty(a, b)$ and $c \in (a, b)$. If for every $x \in (a, b)$, $\lim_{n\to\infty} |R_n(x)| = 0$, then we have*

$$f(x) = f(c) + \frac{f'(c)}{1!}(x-c) + \frac{f''(c)}{2!}(x-c)^2 + \cdots + \frac{f^{(k)}(c)}{k!}(x-c)^k + \cdots$$

$$= \sum_{k=0}^{\infty} \frac{f^{(k)}(c)}{k!}(x-c)^k.$$

Proof. In Definition 7.5.3, we have

$$f(x) = P_n(x) + \frac{f^{(n+1)}(d)}{(n+1)!}(x-c)^{n+1} \quad (c < d < x)$$
$$= P_n(x) + R_n(x).$$

If $\lim_{n\to\infty} |R_n(x)| = 0$, then we have $\lim_{n\to\infty}(f(x) - P_n(x)) = 0$, and thus we have

$$f(x) = \lim_{n\to\infty} P_n(x) = \lim_{n\to\infty} \sum_{k=0}^{n} \frac{f^k(c)}{k!}(x-c)^k = \sum_{k=0}^{\infty} \frac{f^{(k)}(c)}{k!}(x-c)^k.$$

This completes the proof.

Definition 7.5.4. (1) In Theorem 7.5.2,

$$f(x) = \sum_{k=0}^{\infty} \frac{f^{(k)}(c)}{k!}(x-c)^k$$

is called the *Taylor series* for $f(x)$ at c;

(2) If $c = 0$ in the Taylor series, then

$$f(x) = \sum_{k=0}^{\infty} \frac{f^{(k)}(0)}{k!}x^k$$

is called the *Maclaurin series* for $f(x)$.

Example 7.5.2. Prove the following:

(1) $e^x = \sum_{n=0}^{\infty} \frac{1}{n!}x^n$;

(2) $\sin x = \sum_{n=0}^{\infty} \frac{(-1)^n}{(2n+1)!}x^{2n+1}$;

(3) $\cos x = \sum_{n=0}^{\infty} \frac{(-1)^n}{2n!}x^{2n}$.

Solution. (1) Let $f(x) = e^x$. Then, for every $n \geq 1$, $f^{(n)}(x) = e^x$ and so, by using the Taylor Theorem at $c = o$, for $0 < d < x$ or $x < d < 0$, we have

$$f(x) = f(0) + \frac{f'(0)}{1!}x + \frac{f''(0)}{2!}x^2 + \cdots + \frac{f^{(n)}(0)}{n!}x^n + \frac{f^{(n+1)}(d)}{(n+1)!}x^{n+1}$$
$$= 1 + x + \frac{x^2}{2!} + \cdots + \frac{x^n}{n!} + \frac{f^{(n+1)}(d)}{(n+1)!}x^{n+1}$$
$$= 1 + x + \frac{x^2}{2!} + \cdots + \frac{x^n}{n!} + R_n(x).$$

Now, we have

$$\lim_{n\to\infty} |R_n(x)| = \lim_{n\to\infty} \left| \frac{x^{n+1}}{(n+1)!} \right| |e^d| = e^d \lim_{n\to\infty} \frac{x^{n+1}}{(n+1)!} = 0.$$

Therefore, we have

$$e^x = 1 + x + \frac{x^2}{2!} + \cdots + \frac{x^n}{n!} + \cdots = \sum_{n=0}^{\infty} \frac{x^n}{n!}.$$

(2) Let $f(x) = \sin x$. Then, we have

$$f'(x) = \cos x, \ \ f''(x) = -\sin x, \ \ f'''(x) = -\cos x, \ \ \ldots$$

and

$$f^{(n)}(x) = \begin{cases} \cos x, & \text{if } n = 1, 5, 9, \ldots, \\ -\sin x, & \text{if } n = 2, 6, 10, \ldots, \\ -\cos x, & \text{if } n = 3, 7, 11, \ldots, \\ \sin x, & \text{if } n = 0, 4, 8, 12, \ldots \end{cases}$$

and so

$$f^{(n)}(0) = \begin{cases} 1, & \text{if } n = 1, 5, 9, \ldots, \\ -1, & \text{if } n = 3, 7, 11, \ldots, \\ 0, & \text{otherwise.} \end{cases}$$

Thus, we have

$$\lim_{n \to \infty} |R_n(x)| = \lim_{n \to \infty} \left| \frac{f^{(n)}(d)}{(n+1)!} x^{N+1} \right| \leq \frac{|x|^{n+1}}{(n+1)!}$$

and so, since $\lim_{n \to \infty} \frac{|x|^{n+1}}{(n+1)!} = 0$,

$$\lim_{n \to \infty} |R_n(x)| = 0.$$

Therefore, we have

$$\sin x = \sin 0 + \frac{\cos 0}{1!} x - \frac{\sin 0}{2!} x^2 + \cdots + \frac{f^{(n)}(0)}{n!} + \cdots$$

$$= x - \frac{x^3}{3!} + \frac{x^5}{5!} + \cdots + (-1)^n \frac{x^{2n+1}}{(2n+1)!} + \cdots$$

$$= \sum_{n=0}^{\infty} (-1)^n \frac{x^{2n+1}}{(2n+1)!}.$$

(3) Since $\cos x = (\sin x)'$, it follows from (2) that

$$\cos x = 1 - \frac{x^2}{2!} + \frac{x^4}{4!} + \cdots + (-1)^n \frac{x^{2n}}{2n!}$$

$$= \sum_{n=0}^{\infty} \frac{(-1)^n}{2n!} x^{2n}.$$

Example 7.5.3. Compute the value of sin 47° accurate to four decimal places.

Solution. The Taylor series of $\sin x$ at c is as follows:

$$\sin x = \sin c + \cos c(x - c) - \sin c \frac{(x - c)^2}{2!}$$

$$- \cos c \frac{(x - c)^3}{3!} + \cdots . \qquad (7.5.4)$$

To make $(x - c)$ small, we have to choose a value of c near to the value of x for which the function value is being computed. $\sin c$ and $\cos c$ have to be known. Therefore, we choose $c = \frac{1}{4}\pi$, and then, from (7.5.4), we have

$$\sin x = \sin \frac{1}{4}\pi + \cos \frac{1}{4}\pi \left(x - \frac{1}{4}\pi\right) - \sin \frac{1}{4}\pi \frac{(x - \frac{1}{4}\pi)^2}{2!}$$

$$- \cos \frac{1}{4}\pi \frac{(x - \frac{1}{4}\pi)^3}{3!} + \cdots . \qquad (7.5.5)$$

Since $47° = \frac{47}{180}\pi = (\frac{1}{4}\pi + \frac{1}{90}\pi)$, it follows from (7.5.5) that

$$\sin \frac{47}{180}\pi = \frac{1}{2}\sqrt{2} + \frac{1}{2}\sqrt{2} \cdot \frac{1}{90}\pi - \frac{1}{2}\sqrt{2} \cdot \frac{1}{2}\left(\frac{1}{90}\pi\right)^2$$

$$- \frac{1}{2}\sqrt{2} \cdot \frac{1}{6}\left(\frac{1}{90}\pi\right)^3 + \cdots \qquad (7.5.6)$$

$$\approx \frac{1}{2}\sqrt{2}(1 + 0.03490 - 0.00061 - 0.000002 + \cdots).$$

Thus, take $\sqrt{2} = 1.4142$ and using the first three terms of (7.5.6), we have

$$\sin \frac{47}{180}\pi \approx (0.70711)(1.03490) = 0.73136,$$

which implies that

$$\sin 47° \approx 0.73136$$

and, further,

$$\left|R_2\left(\frac{47}{180}\pi\right)\right| \leq \frac{\left(\frac{1}{90}\pi\right)^3}{3!} \approx 0.00001.$$

Therefore, the result is accurate to four decimal places.

By using Taylor's Theorem, we have the following:

Theorem 7.5.3. *Let $f \in C^\infty(a, b)$ and $c \in (a, b)$ be such that*

$$f'(c) = f''(c) = \cdots = f^{(n-1)}(c) = 0, \quad f^{(n)}(c) \neq 0.$$

Then, we have the following:

(1) *If n is an even positive integer and $f^{(n)}(c) > 0$, then f has the local minimum at c;*

(2) *If n is an even positive integer and $f^{(n)}(c) < 0$, then f has the local maximum at c;*

(3) *If n is an odd positive integer, then f has not the local minimum and maximum at c.*

Proof. Let $P_n(x)$ be the Taylor polynomial of degree n of f at c, that is,

$$P_n(x) = f(c) + \frac{f'(c)}{1!}(x - c) + \frac{f''(c)}{2!}(x - c)^2 + \cdots + \frac{f^{(n)}(c)}{n!}(x - c)^n.$$

Then, for every $x \in (a, b)$, there exists $d \in (c, x)$ such that

$$f(x) = P_{n-1}(x) + R_{n-1}(x) = f(c) + \frac{f^{(n)}(d)}{n!}(x - c)^n. \tag{7.5.7}$$

Since $f^{(n)}$ is continuous, it follows that if $x \to c$, then we have

$$f^{(n)}(x) \to f^{(n)}(c).$$

So, since $f^{(n)}(c) \neq 0$, we can choose any $\varepsilon > 0$ such that

$$0 < \varepsilon < |f^{(n)}(c)|$$

and so there exists $\delta > 0$ such that if $|x - c| < \delta$, then

$$|f^{(n)}(x) - f^{(n)}(c)| < \varepsilon.$$

Therefore, for every $x \in (c - \delta, c + \delta)$, if $f^{(n)}(c) > 0$, then $f^{(n)}(x) > 0$, and, if $f^{(n)}(c) < 0$, then $f^{(n)}(x) < 0$.

(1) Let n be an even positive integer and $f^{(n)}(c) > 0$. Then, for every $x \in (c - \delta, c + \delta) \setminus \{c\}$, since $f^{(n)}(d) > 0$ and $(x - c)^n > 0$, it follows from (7.5.7) that

$$f(x) - f(c) = \frac{f^{(n)}(d)}{n!}(x - c)^n > 0.$$

Thus, for every $x \in (c - \delta, c + \delta) \setminus \{c\}$, we have $f(x) > f(c)$ and so f has the local minimum at c.

(2) Let n be an even positive integer and $f^{(n)}(c) < 0$. Then, for every $x \in (c - \delta, c + \delta) \setminus \{c\}$, since $f^{(n)}(d) < 0$ and $(x - c)^n > 0$, it follows from (7.5.7) that

$$f(x) - f(c) = \frac{f^{(n)}(d)}{n!}(x - c)^n < 0.$$

Thus, for every $x \in (c - \delta, c + \delta) \setminus \{c\}$, we have $f(x) < f(c)$ and so f has the local maximum at c.

(3) Let n be an odd positive integer. If $x \in (c - \delta, c)$, then $(x - c)^n < 0$, and if $x \in (c, c + \delta)$, then $(x - c)^n > 0$. Let $\eta > 0$. For any $x \in (c - \eta, c + \eta)$, it follows that both $f(x) > f(c)$ and $f(x) < f(c)$ do not hold. Therefore, f has not the local minimum and maximum at c.

This completes the proof.

7.6   Exercises

Excercise 7.1 Show that the following function

$$f(x) = \begin{cases} x^2 \sin \frac{1}{x}, & \text{if } x \neq 0, \\ 0, & \text{if } x = 0, \end{cases}$$

is differentiable at $x = 0$. Also, verify if $\lim_{x \to 0} f'(x) = f'(0)$. Finally, draw a conclusion from your verification.

Excercise 7.2 Discuss the continuity and differentiability of the following function

$$f(x) = |x - 2| + |x - 3|$$

on the interval $[0, 4]$.

Excercise 7.3 Show that the following function

$$f(x) = \begin{cases} x \frac{e^{\frac{1}{x}} - e^{-\frac{1}{x}}}{e^{\frac{1}{x}} + e^{-\frac{1}{x}}}, & \text{if } x \neq 0, \\ 0, & \text{if } x = 0, \end{cases}$$

is continuous but not differentiable at $x = 0$.

Excercise 7.4 Suppose that $h(x)$ is the inverse function for $f(x) = 3x^4 + 5x^2 + 6$. Find the value of $h'(12)$.

Excercise 7.5 Use the inverse function theorem to find the derivative of $f(x) = \sqrt[3]{x}$.

Hint. The function $f(x) = \sqrt[3]{x}$ is the inverse of the function $g(x) = x^3$.

Excercise 7.6 If p is an integer and q is an arbitrary integer, then prove that

$$\frac{d}{dx}(x^{\frac{q}{p}}) = \frac{q}{p} x^{(\frac{q}{p})-1}.$$

Hint. The function $f(x) = x^{\frac{1}{p}}$ is the inverse of the function $g(x) = x^p$. Then, use inverse function theorem, and apply chain rule.

Excercise 7.7 Find the derivative of $\cos^{-1}(2x - 1)$.

Excercise 7.8 Show that the following function

$$f(x) = x^2 - 2x + 5$$

meets the criteria for Rolle's Theorem on the interval $[0, 2]$. Then, find the point where $f'(x) = 0$.

Excercise 7.9 Consider the following function:

$$f(x) = \begin{cases} x + 1, & \text{if } x \leq 2, \\ -x^2 + 4x + 2, & \text{if } x > 2. \end{cases}$$

There is no point $x_0 \in (1, 4)$, where $f'(x_0) = 0$. Why doesn't Rolle's Theorem apply to this situation?

Hint. $f(1) = 2 = f(4)$, but $f(x)$ is not continuous at every $x \in [1, 4]$.

Excercise 7.10 Check the validity of Lagrange's Mean Value Theorem for the following function

$$f(x) = x^2 - 3x + 5$$

on $[1, 3]$. If the theorem holds, then find a point ξ satisfying the conditions of the theorem.

Ans. $\xi = 2$.

Excercise 7.11 If a function is everywhere continuous and differentiable and has two real roots, then its derivative has at least one root.

Hint. Use Lagrange's Mean Value Theorem.

Excercise 7.12 Examine the validity of Cauchy's Mean Value Theorem for the functions $\sin x$ and $\cos x$ on $[a, b]$.

Excercise 7.13 Show that

(1) $\cos x > 1 - \frac{x^2}{2}$ for every $x \neq 0$;

(2) $\frac{\cos x - \cos y}{\sin y - \sin x} = \tan \theta$ for every $0 < y < \theta < x < \frac{\pi}{2}$.

Hint. (1) Use Cauchy's Mean Value Theorem for $1 - \cos x$ and $\frac{x^2}{2}$;

(2) Use Cauchy's Mean Value Theorem for $\sin x$ and $\cos x$.

Excercise 7.14 Evaluate each of the following limits:

(1) $\lim\limits_{x \to 0} \frac{\sin x}{x}$;

(2) $\lim\limits_{x \to \infty} \frac{e^x}{x^2}$;

(3) $\lim\limits_{x \to \infty} x^{\frac{1}{x}}$;

(4) $\lim\limits_{x \to 0} \frac{x - \sin x}{x^3}$;

(5) $\lim\limits_{x \to \infty} \frac{e^x + 3x^2}{7x + e^x}$.

Ans. (1) 1, (2) ∞, (3) 1, (4) $\frac{1}{6}$, (5) 1.

Excercise 7.15 Find the following:

(1) $\lim\limits_{x \to 1+} \left(\frac{1}{x-1} - \frac{1}{\ln x} \right)$;

(2) $\lim\limits_{x \to \infty} \frac{\sqrt{5x+7}}{\sqrt{x+8}}$;

(3) $\lim\limits_{x \to -\infty} xe^x$;

(4) $\lim\limits_{x \to 0+} \sin x \log x$;

(5) $\lim\limits_{x \to 0+} x^{x^x}$.

Excercise 7.16 Prove the following:

(1) $\ln(1 + x) = x - \frac{x^2}{2} + \frac{x^3}{3} - \frac{x^4}{4} \cdots$;

(2) $e^{-2x} = 1 - 2x + 2x^2 - \frac{4}{3}x^3 + \cdots$.

Hint. Use Taylor's Theorem.

Excercise 7.17 Find the Taylor series for $\frac{1}{x^2}$ about $x = -3$.

Excercise 7.18 Explain why the function $\frac{1}{x}$ has no local maxima or minima.

Excercise 7.19 Find all local maximum and minimum points for the function $4x^3 - 3x$.

Excercise 7.20 Find the maximum value of $\left(\frac{1}{x} \right)^x$.

Ans. $e^{\frac{1}{e}}$.

Excercise 7.21 Find the maximum value of the function $x^2 e^{-2x}$ for every $x > 0$.

8

Integration

In this chapter, we deal with the Riemann integral, examples and properties of Riemann integral, the Fundamental Theorems of Calculus, the Mean Value Theorems for integrals, the Substitution Theorem for integrals, Theorems on the Integration by Parts, improper Riemann integrals, the Riemann-Stieltjes integral with examples and properties, and functions of bounded variation. Solved examples and chapter-end exercises are also incorporated associating each topic discussed.

8.1   The Riemann Integral

Now, we give the definition of the Riemann integral of a function.

Definition 8.1.1. Let $[a, b]$ be a closed interval.

(1) The set $P = \{x_0, x_1, x_2, \ldots, x_n\}$ of finite points $a = x_0, x_1, x_2, \ldots, x_n = b$ with

$$a = x_0 < x_1 < x_2 < \cdots < x_n = b$$

is called a *partition* of $[a, b]$;

(2) For every $i = 1, 2, \ldots, n$, a closed interval $I_i = [x_{i-1}, x_i]$ is called the *subinterval* of $[a, b]$;

(3) For every $i = 1, 2, \ldots, n$, the *length* of $I_i = [x_{i-1}, x_i]$ is defined by $\Delta x_i = x_i - x_{i-1}$;

(4) $\wp[a, b]$ is denoted by the set of all partitions of $[a, b]$.

Definition 8.1.2. Let $f : [a, b] \to \mathbb{R}$ be a bounded function and $P = \{a = x_0, x_1, x_2, \ldots, x_n = b\}$ be a partition of $[a, b]$. For every subinterval $I_i = [x_{i-1}, x_i]$, put

$$M_i = \sup\{f(x) : x \in [x_{i-1}, x_i]\}, \quad m_i = \inf\{f(x) : x \in [x_{i-1}, x_i]\}.$$

Then, for every $P \in \wp[a, b]$, define the *upper Riemann sum* $U(f, P)$ and the *lower Riemann sum* $L(f, P)$ as follows, respectively:

$$U(f, P) = \sum_{i=1}^{n} M_i \Delta x_i, \quad L(f, P) = \sum_{i=1}^{n} m_i \Delta x_i.$$

Example 8.1.1. Let $f : [a, b] \to \mathbb{R}$ be a function defined by $f(x) = c$, where $c \in \mathbb{R}$ is a constant. Prove that, for every $P = \{a = x_0, x_1, x_2, \ldots, x_n = b\} \in \wp[a, b]$,

$$U(f, P) = L(f, P) = c.$$

Solution. Let $\Delta x_i = x_i - x_{i-1}$ for every $i = 1, 2, \ldots, n$. Since $M_i = m_i = c$, we have the following:

$$U(f, P) = \sum_{i=1}^{n} M_i \Delta x_i = \sum_{i=1}^{n} c \Delta x_i = c \sum_{i=1}^{n} \Delta x_i = c(b - a)$$

and

$$L(f, P) = \sum_{i=1}^{n} m_i \Delta x_i = \sum_{i=1}^{n} c \Delta x_i = c \sum_{i=1}^{n} \Delta x_i = c(b - a).$$

Thus, we have

$$U(f, P) = L(f, P) = c.$$

Example 8.1.2. Let $f : [0, 1] \to \mathbb{R}$ be a function defined by

$$f(x) = \begin{cases} 1, & \text{if } x \in \mathbb{Q}, \\ 0, & \text{if } x \notin \mathbb{Q}^c. \end{cases}$$

Then, calculate $U(f, P)$ and $L(f, P)$.

Solution. Let $P = \{0 = x_0, x_1, x_2, \ldots, x_n = 1\} \in \wp[0, 1]$. Each subinterval $I_i = [x_{i-1}, x_i]$ for every $i = 1, 2, \ldots, n$ has infinitely many rational numbers and irrational numbers, and so, for every $i = 1, 2, \ldots, n$,

$$M_i = \sup\{f(x) : x \in [x_{i-1}, x_i]\} = 1$$

and

$$m_i = \inf\{f(x) : x \in [x_{i-1}, x_i]\} = 0.$$

Hence, we have $L(f, P) = 0$ and $U(f, P) = 1$.

From Definition 8.1.2, we have the following.

Lemma 8.1.1. *Let $f : [a, b] \to \mathbb{R}$ be a bounded function and $P = \{a = x_0, x_1, x_2, \ldots, x_n = b\}$ be a partition of $[a, b]$. Then, we have the following:*

(1) $L(f, P) \leq U(f, P)$;

(2) $m(b - a) \leq L(f, P) \leq U(f, P) \leq M(b - a)$, *where* $m, M \in \mathbb{R}$ *are numbers such that*

$$m \leq f(x) \leq M \quad \text{for every} \quad x \in [a, b].$$

Definition 8.1.3. Let $[a, b]$ be a closed interval.

(1) For every $P, Q \in \wp[a, b]$, Q is called the *refinement* of P if

$$P \subset Q;$$

(2) The partition $R = P \cup Q$ is called the *common refinement* of P and Q.

Lemma 8.1.2. *Let* $f : [a, b] \to \mathbb{R}$ *be a bounded function. For every* $P, Q \in \wp[a, b]$, *if* Q *is the refinement of* P, *then we have the following:*

(1) $L(f, P) \leq L(f, Q)$;

(2) $U(f, Q) \leq U(f, P)$.

Proof. (1) First, let

$$P = \{a = x_0, x_1, x_2, \ldots, x_n = b\}$$

and

$$Q = \{a = x_0, x_1, x_2, \ldots, x_{j-1}, x^*, x_j, \ldots, x_n = b\},$$

that is, $Q = P \cup \{x^*\}$. Then, Q is a refinement of P, that is, $P \subset Q$. Next, put

$$m_j' = \inf\{f(x) : x \in [x_{j-1}, x^*]\}, \quad m_j'' = \inf\{f(x) : x \in [x^*, x_j]\}$$

and

$$m_j = \min\{m_j', m_j''\}.$$

Then, we have

$$L(f, P) = \sum_{i=1}^{n} m_i \Delta x_i = \sum_{i=1, i \neq j}^{n} m_i \Delta x_i + m_j \Delta x_j$$

and

$$L(f, Q) = \sum_{i=1}^{j-1} m_i \Delta x_i + m_j(x^* - x_{j-1}) + m_j(x_j - x^*) + \sum_{i=j+1}^{n} m_i \Delta x_i.$$

Thus, since $m_j \leq m_j', m_j''$, we have

$$m_j \Delta x_j = m_j(x^* - x_{j-1}) + m_j(x_j - x^*)$$
$$\leq m_j'(x^* - x_{j-1}) + m_j''(x_j - x^*)$$

and so

$$\sum_{i=1, i\neq j}^{n} m_j \Delta x_j = \sum_{i=1, i\neq j}^{n} m_j(x^* - x_{j-1}) + \sum_{i=1, i\neq j}^{n} m_j(x_j - x^*)$$

$$\leq \sum_{i=1, i\neq j}^{n} m_j'(x^* - x_{j-1}) + \sum_{i=1, i\neq j}^{n} m_j''(x_j - x^*),$$

which implies that $L(f, P) \leq L(f, Q)$.

Next, let Q be the refinement adjoining a finite points to P, and then, repeating the proceeding process, we can prove $L(f, P) \leq L(f, Q)$.

(2) By the same argument as in (1), we can prove $U(f, Q) \leq U(f, P)$. This completes the proof.

Theorem 8.1.3. *Let $f : [a, b] \to \mathbb{R}$ be a bounded function. For every $P, Q \in \wp[a, b]$, we have*

$$L(f, P) \leq U(f, Q).$$

Proof. Let $T = P \cup Q$. Then, T is a refinement of P and Q. Thus, by Lemma 8.1.2, we have

$$L(f, P) \leq L(f, T), \quad U(f, T) \leq U(f, Q).$$

Therefore, since $L(f, P) \leq U(f, P)$, we have

$$L(f, P) \leq U(f, Q).$$

This completes the proof.

Corollary 8.1.4. *Let $f : [a, b] \to \mathbb{R}$ be a bounded function. For every $P, Q \in \wp[a, b]$, if $L(f, P) \leq U(f, Q)$, then*

$$\sup\{L(f, P) : P \in \wp[a, b]\}, \quad \inf\{U(f, P) : P \in \wp[a, b]\}$$

exist.

Proof. Let $Q \in \wp[a, b]$ be fixed. Then, for every $P \in \wp[a, b]$, we have

$$L(f, P) \leq U(f, Q), \quad L(f, Q) \leq U(f, P),$$

which means that $U(f, Q)$ and $L(f, Q)$ are upper bound and lower bound of the following sets:

$$\{L(f, P) : P \in \wp[a, b]\}, \quad \{U(f, P) : P \in \wp[a, b]\},$$

respectively. Therefore, by the Completeness Axiom of $\mathbb{R}$, it follows that

$$\sup\{L(f, P) : P \in \wp[a, b]\}, \quad \inf\{U(f, P) : P \in \wp[a, b]\}$$

exist. This completes the proof.

Corollary 8.1.5. *Let $f, g : [a, b] \to \mathbb{R}$ be bounded functions and $c \in \mathbb{R}$. For every $P \in \wp[a, b]$, we have the following:*

(1) $U(f + g, P) \leq U(f, P) + U(g, P)$;

(2) $L(f + g, P) \geq L(f, P) + L(g, P)$;

(3) *If $c \geq 0$, then we have*

$$U(cf, P) = cU(f, P), \quad L(cf, P) = cL(f, P);$$

(4) *If $c < 0$, then we have*

$$U(cf, P) = cL(f, P), \quad L(cf, P) = cU(f, P).$$

Proof. Let $P = \{a = x_0, x_1, x_2, \ldots, x_n = b\} \in \wp[a, b]$.

(1) We have

$$U(f + g, P) = \sum_{i=1}^{n} \sup\{f(x) + g(x) : x \in [x_{i-1}, x_i]\} \Delta x_i$$

$$\leq \sum_{i=1}^{n} [\sup\{f(x) : x \in [x_{i-1}, x_i]\}$$

$$+ \sup\{g(x) : x \in [x_{i-1}, x_i]\}] \Delta x_i$$

$$= U(f, P) + U(g, P).$$

(2) As in the proof of (1), we have $L(f + g, P) \geq L(f, P) + L(g, P)$.

(3) Let $c \geq 0$. Then, we have

$$U(cf, P) = \sum_{i=1}^{n} \sup\{cf(x) : x \in [x_{i-1}, x_i]\} \Delta x_i$$

$$= c \sum_{i=1}^{n} \sup\{f(x) : x \in [x_{i-1}, x_i]\} \Delta x_i$$

$$= cU(f, P).$$

(4) Let $c < 0$. Then, we have

$$U(cf, P) = \sum_{i=1}^{n} \sup\{cf(x) : x \in [x_{i-1}, x_i]\} \Delta x_i$$

$$= c \inf_{i=1}^{n} \sup\{f(x) : x \in [x_{i-1}, x_i]\} \Delta x_i$$

$$= cL(f, P).$$

This completes the proof.

Definition 8.1.4. Let $f : [a, b] \to \mathbb{R}$ be a bounded function.

(1) The *upper Riemann integral* of f on $[a, b]$ is the number

$$\overline{\int_a^b} f(x)dx = \inf\{U(f, P) : P \in \wp[a, b]\};$$

(2) The *lower Riemann integral* of f on $[a, b]$ is the number

$$\underline{\int_a^b} f(x)dx = \sup\{L(f, P) : P \in \wp[a, b]\};$$

(3) f is said to be *Riemann integrable* on $[a, b]$ if

$$\overline{\int_a^b} f(x)dx = \underline{\int_a^b} f(x)dx;$$

(4) In this case, the *Riemann integral* of f is defined to be the number

$$\overline{\int_a^b} f(x)dx = \underline{\int_a^b} f(x)dx,$$

which is denoted by

$$\int_a^b f(x)dx;$$

(5) In addition, we define

$$\int_a^b f(x)dx = -\int_b^a f(x)dx, \quad \int_a^a f(x)dx = 0;$$

(6) f is said to be *not Riemann integrable* on $[a, b]$ if

$$\overline{\int_a^b} f(x)dx \neq \underline{\int_a^b} f(x)dx.$$

Note that $\Re[a, b]$ denotes the set of Riemann integrable functions on $[a, b]$.

From the definition of the Riemann integral, it follows that, for every $p, Q \in \wp[a, b]$ with $P \subset Q$,

$$L(f, P) \leq L(f, Q) \leq \int_a^b f(x)dx \leq U(f, Q) \leq U(f, P).$$

Example 8.1.4. Let $f : [a, b] \to \mathbb{R}$ be the function defined by

$$f(x) = \begin{cases} 1, & \text{if } x \in \mathbb{Q}, \\ 0, & \text{if } x \notin \mathbb{Q}^c. \end{cases}$$

Then, we have $L(f, P) = 0$ and $U(f, P) = 1$ and so

$$\overline{\int_a^b} f(x)dx = \inf\{U(f, P) : P \in \wp[a, b]\} = 1$$

and

$$\underline{\int_a^b} f(x)dx = \sup\{L(f, P) : P \in \wp[a, b]\} = 0.$$

Therefore, sine $\overline{\int_a^b} f(x)dx = 1 \neq 0 = \underline{\int_a^b} f(x)dx$, f is not Riemann integrable on $[0, 1]$.

Example 8.1.5. Let $f : [a, b] \to \mathbb{R}$ be the function defined by $f(x) = x$ for every $x \in [a, b]$. By using the definition of the Riemann integral, show that

$$\int_a^b f(x)dx = \frac{1}{2}(b^2 - a^2).$$

Solution. For every $n \geq 1$, let P_n be a partition of $[a, b]$ given by

$$P_n = \left\{a, a + \frac{b-a}{n}, a + \frac{2(b-a)}{n}, \dots, a + \frac{n(b-a)}{n} = b\right\}.$$

Since $f(x) = x$ is increasing on $a, b]$, at each subinterval $\left[a + \frac{(i-1)(b-a)}{n}, a + \frac{i(b-a)}{n}\right]$, f has the minimum value $m_i = a + \frac{(i-1)(b-a)}{n}$ and the maximum value $M_i = a + \frac{i(b-a)}{n}$. Therefore, we have

$$L(f, P_n) = \sum_{i=1}^n m_i \Delta x_i$$

$$= \sum_{i=1}^n \left(a + \frac{(i-1)(b-a)}{n}\right)\left(\frac{b-a}{n}\right)$$

$$= a(b-a) + \frac{(b-a)^2}{2}\left(1 - \frac{1}{n}\right).$$

Similarly, we have

$$L(f, P_n) = \sum_{i=1}^n M_i \Delta x_i$$

$$= \sum_{i=1}^n \left(a + \frac{i(b-a)}{n}\right)\left(\frac{b-a}{n}\right)$$

$$= a(b-a) + \frac{(b-a)^2}{2}\left(1 + \frac{1}{n}\right).$$

If $n \leq m$, then we have

$$L(f, P_n) \leq L(f, P_m), \quad U(f, P_m) \leq U(f, P_n)$$

and so $\{L(f, P_n)\}$ is an increasing sequence and $\{U(f, P_n)\}$ is a decreasing sequence. Thus, we have

$$\sup_{n \geq 1} L(f, P_n) = \lim_{n \to \infty} L(f, P_n) = \frac{b^2 - a^2}{2}$$

and

$$\inf_{n \geq 1} U(f, P_n) = \lim_{n \to \infty} U(f, P_n) = \frac{b^2 - a^2}{2}.$$

On the other hand, we have

$$\sup_{n \geq 1} L(f, P_n) \leq \sup\{L(f, P) : P \in \wp[a, b]\} = \int_{\underline{a}}^{b} f(x)dx$$

and

$$\inf_{n \geq 1} L(f, P_n) \geq \inf\{L(f, P) : P \in \wp[a, b]\} = \overline{\int_{a}^{b} f(x)dx}$$

and so

$$\frac{b^2 - a^2}{2} \leq \int_{\underline{a}}^{b} f(x)dx \leq \overline{\int_{a}^{b} f(x)dx} = \frac{b^2 - a^2}{2}.$$

Thus, we have

$$\int_{\underline{a}}^{b} f(x)dx = \overline{\int_{a}^{b} f(x)dx} = \frac{b^2 - a^2}{2},$$

which implies that $f(x) = x$ is Riemann integrable on $[a, b]$ and

$$\int_{a}^{b} f(x)dx = \frac{1}{2}(b^2 - a^2).$$

Theorem 8.1.6. (The Riemann Integral Test) *Let* $f : [a, b] \to \mathbb{R}$ *be a bounded function. Then, the following are equivalent:*

(1) $f \in \Re[a, b]$;

(2) *For every* $\varepsilon > 0$, *there exists* $P \in \wp[a, b]$ *such that*

$$U(f, P) - L(f, P) < \varepsilon.$$

Proof. (1) $\implies$ (2) Suppose that $f \in \Re[a, b]$. Since

$$\int_{a}^{b} f(x)dx = \inf\{U(f, P) : P \in \wp[a, b]\},$$

for every $\varepsilon > 0$, $\int_{a}^{b} f(x)dx + \frac{\varepsilon}{2}$ is not a lower bound of $\{U(f, P) : P \in \wp[a, b]\}$, and so there exists $P_1 \in \wp[a, b]$ such that

$$U(f, P_1) < \int_{a}^{b} f(x)dx + \frac{\varepsilon}{2}.$$

Also, since

$$\int_a^b f(x)dx = \inf\{L(f, P) : P \in \wp[a, b]\},$$

$\int_a^b f(x)dx - \frac{\varepsilon}{2}$ is not an upper bound of $\{L(f, P) : P \in \wp[a, b]\}$, and so there exists $P_2 \in \wp[a, b]$ such that

$$\int_a^b f(x)dx - \frac{\varepsilon}{2} < L(f, P_2).$$

Letting $P = P_1 \cup P_2$, we have

$$L(f, P_2) \le L(f, P) \le U(f, P) \le U(f, P_1).$$

Thus, we have

$$U(f, P) - L(f, P) \le U(f, P_1) - L(f, P_2)$$
$$< \left(\int_a^b f(x)dx + \frac{\varepsilon}{2}\right) - \left(\int_a^b f(x)dx - \frac{\varepsilon}{2}\right)$$
$$= \varepsilon.$$

$(2) \Longrightarrow (1)$ Suppose that, for every $\varepsilon > 0$, there exists $P \in \wp[a, b]$ such that

$$U(f, P) - L(f, P) < \varepsilon.$$

Since $\overline{\int_a^b} f(x)dx \le U(f, P)$ and $L(f, P) \le \underline{\int_a^b} f(x)dx$, we have

$$0 \le \overline{\int_a^b} f(x)dx - \underline{\int_a^b} f(x)dx \le U(f, P) - L(f, P) < \varepsilon$$

and so

$$\left|\overline{\int_a^b} f(x)dx - \underline{\int_a^b} f(x)dx\right| = \overline{\int_a^b} f(x)dx - \underline{\int_a^b} f(x)dx < \varepsilon.$$

Therefore, we have

$$\overline{\int_a^b} f(x)dx = \underline{\int_a^b} f(x)dx$$

and $f \in \Re[a, b]$. This completes the proof.

From Theorem 8.1.6, we have the following:

Corollary 8.1.7. *Let $f : [a, b] \to \mathbb{R}$ be a bounded function. Then, the following are equivalent:*

(1) $f \in \Re[a, b]$;

(2) *There exists a sequence $\{P_n\}$ of partitions of $[a, b]$*

$$\lim_{n\to\infty} (U(f, P_n) - L(f, P_n)) = 0.$$

In this case, we have

$$\lim_{n\to\infty} L(f, P_n) = \int_a^b f(x)dx = \lim_{n\to\infty} U(f, P_n).$$

Example 8.1.6. Let $f : [0, 1] \to \mathbb{R}$ be the function defined by $f(x) = x$ for every $x \in [0, 1]$. By using the Riemann Integral Test, prove that $f(x) = x$ is Riemann integrable on $[0, 1]$.

Solution. As in Example 8.1.5, let $P_n \in \wp[0, 1]$ be a partition of $[a, b]$. Then, we have

$$L(f, P_n) = \frac{n-1}{2n}, \quad U(f, P_n) = \frac{n+1}{2n}.$$

For every $\varepsilon > 0$, if we take a positive integer n_0 such that

$$\frac{1}{\varepsilon} < n_0,$$

then we have

$$U(f, P_{n_0}) - L(f, P_{n_0}) = \frac{1}{n_0} < \varepsilon.$$

Therefore, by Theorem 8.1.6 (the Riemann Integral Test), f is Riemann integrable on $[0, 1]$.

A function $f : [a, b] \to \mathbb{R}$ is said to be *monotone* on $[a, b]$ if it is either increasing or decreasing on $[a, b]$.

Theorem 8.1.8. *Every monotone function $f : [a, b] \to \mathbb{R}$ is Riemann integrable.*

Proof. Assume that f is increasing on $[a, b]$. Since $f(a) \leq f(x) \leq f(b)$ for every $x \in [a, b]$, f is clearly bounded on $[a, b]$. Let $\varepsilon > 0$, and select a positive integer n_0 such that

$$\frac{(f(b) - f(a))(b - a)}{n} < \varepsilon.$$

For a partition

$$P = \{a = x_0 < x_1 < \cdots < x_{n-1} < x_n = b\},$$

where $\Delta x_i = x_i - x_{i-1} = \frac{b-a}{n}$ for every $i \geq 1$, we have

$$U(f, P) = \sum_{i=1}^{n} M_i \Delta x_i = \sum_{i=1}^{n} f(x_i) \frac{b-a}{n}$$

and

$$L(f, P) = \sum_{i=1}^{n} m_i \Delta x_i = \sum_{i=1}^{n} f(x_{i-1}) \frac{b-a}{n}.$$

Thus, since f is increasing on $[a, b]$, we have

$$U(f, P) - L(f, P) = \frac{b-a}{n} \sum_{i=1}^{n} |f(x_i) - f(x_{i-1})|$$

$$= \frac{b-a}{n} (f(b) - f(a))$$

$$< \varepsilon.$$

Therefore, by the Riemann Integral Test, f is Riemann integrable on $[a, b]$.

Similarly, we can prove this theorem when f is decreasing on $[a, b]$. This completes the proof.

Theorem 8.1.9. *Let $f : [a, b] \rightarrow \mathbb{R}$ be a continuous function. Then, $f \in \Re[a, b]$, that is, f is Riemann integrable on $[a, b]$.*

Proof. Since f is a continuous function on $[a, b]$, f is uniformly continuous on $[a, b]$. Thus, for every $\varepsilon > 0$, there exists $\delta > 0$ such that

$$|x - y| < \delta, \; x, y \in [a, b] \implies |f(x) - f(y)| < \frac{\varepsilon}{b-a}.$$

Now, select a sufficiently large positive integer n_0, and consider a partition $P = \{x_0, x_1, x_2, \ldots, x_{n_0}\}$ of $[a, b]$ with

$$\Delta x_1 = \Delta x_2 = \cdots = \Delta x_{n_0} = \frac{b-a}{n_0}.$$

Further, for each subinterval $I_i = [x_{i-1}, x_i]$ of the partition P_{n_0} $(1 \leq i \leq n_0)$, there exist $t_i, u_i \in I_i$ such that

$$M_i = \sup\{f(x) : x \in [x_{i-1}, x_i]\} = f(t_i)$$

and

$$m_i = \inf\{f(x) : x \in [x_{i-1}, x_i]\} = f(u_i).$$

Thus, we have

$$0 \leq U(f, P_{n_0}) - L(f, P_{n_0})$$

$$= \sum_{i=1}^{n_0} (M_i - m_i) \Delta x_i$$

$$= \sum_{i=1}^{n_0} (f(t_i) - f(u_i)) \Delta x_i$$

$$= \sum_{i=1}^{n_0} |f(t_i) - f(u_i)| \left(\frac{b-a}{n_0} \right)$$

$$< \frac{\varepsilon}{b-a} (b-a)$$

$$= \varepsilon.$$

Therefore, by the Riemann Integral Test, f is Riemann integrable on $[a, b]$. This completes the proof.

8.2   Properties of the Riemann Integral

In this section, we give the Riemann integrals of the composition and the product of two Riemann integrable functions and some basic properties of the Riemann integral including some algebraic properties of the Riemann integral.

Theorem 8.2.1. *Let $f : [a, b] \rightarrow \mathbb{R}$ be Riemann integrable on $[a, b]$ and $g : [c, d] \rightarrow \mathbb{R}$ be a continuous function on $[c, d]$ with $f([a, b]) \subset [c, d]$. Then, the composition $g \circ f$ is Riemann integrable on $[a, b]$.*

Proof. Let $\varepsilon > 0$. Since g is continuous on $[c, d]$, let

$$M = \sup\{f(x) : x \in [c, d]\}$$

and let

$$\varepsilon' = \frac{\varepsilon}{2M + (b-a)}.$$

Also, g is uniformly continuous on $[c, d]$, and there exists $\delta > 0$ such that $0 < \delta < \varepsilon'$ and

$$y, z \in [c, d], \ |y - z| \leq \delta \implies |g(y) - g(z)| < \varepsilon'.$$

On the other hand, since f is Riemann integrable on $[a, b]$, there exists a partition $P = \{a = x_0, x_1, x_2, \ldots, x_n = b\}$ of $[a, b]$ such that

$$U(f, P) - L(f, P) < \delta^2.$$

Now, we show that, for this partition P,

$$U(g \circ f, P) - L(g \circ f, P) \leq \varepsilon$$

and then $g \circ f$ is Riemann integrable on $[a, b]$. For every $i = 1, 2, \ldots, n$, put

$$M_i = \sup\{f(x) : x \in [x_{i-1}, x_i]\},$$
$$m_i = \inf\{f(x) : x \in [x_{i-1}, x_i]\},$$
$$M_i' = \sup\{(g \circ f)(x) : x \in [x_{i-1}, x_i]\},$$
$$m_i' = \inf\{(g \circ f)(x) : x \in [x_{i-1}, x_i]\}.$$

Now, to be convenient, we separate the indices of the partition P into two disjoint subsets X and Y as follows:

$$X = \{k : M_i - m_i < \delta\}, \quad Y = \{i : M_i - m_i \geq \delta\}.$$

Now, if $i \in X$, that is, $M_i - m_i < \delta$, then, for every $x, x' \in [x_{i-1}, x_i]$, we have

$$|f(x) - f(x')| < \delta$$

and so

$$|(g \circ f)(x) - (g \circ f)(x')| < \varepsilon',$$

which implies that

$$M_i' - m_i' \leq \varepsilon.$$

Thus, we have

$$\sum_{i \in X}(M_i' - m_i')(x_i - x_{i-1}) \leq \varepsilon'(b - a). \tag{8.2.1}$$

But, if $i \in Y$, then we have $\delta \leq M_i - m_i$ and so

$$\sum_{i \in Y}(x_i - x_{i-1}) \leq \frac{1}{\delta}\sum_{i \in Y}(M_i - m_i)(x_i - x_{i-1})$$
$$\leq \frac{1}{\delta}(U(f, P) - L(f, P))$$
$$< \delta < \varepsilon'.$$

Therefore, we have

$$\sum_{i \in Y}(M_i' - m_i')(x_i - x_{i-1}) \leq 2M\varepsilon'. \tag{8.2.2}$$

If we combine (8.2.1) and (8.2.2), then we have

$$U(g \circ f, P) - L(g \circ f, P)$$
$$= \sum_{i \in X}(M_i' - m_i')(x_i - x_{i-1}) + \sum_{i \in Y}(M_i' - m_i')(x_i - x_{i-1})$$
$$\leq \varepsilon'(b - a) + 2M\varepsilon'$$
$$= \varepsilon.$$

Therefore, by the Riemann Integral Test, $g \circ f$ is Riemann integrable on $[a, b]$. This completes the proof.

Theorem 8.2.2. *Let $f, g : [a, b] \to \mathbb{R}$ be Riemann integrable on $[a, b]$, that is, $f, g \in \mathfrak{R}[a, b]$, and let $c \in \mathbb{R}$. Then, we have the following:*

(1) $f + g \in \mathfrak{R}[a, b]$ *and*

$$\int_a^b (f + g)(x)dx = \int_a^b f(x)dx + \int_a^b g(x)dx;$$

(2) $cf \in \mathfrak{R}[a, b]$ *and*

$$\int_a^b cf(x)dx = c \int_a^b f(x)dx;$$

(3) $fg \in \mathfrak{R}[a, b]$;

(4) *If $f(x) \leq g(x)$ for every $x \in [a, b]$, then*

$$\int_a^b f(x)dx \leq \int_a^b g(x)dx;$$

(5) $|f| \in \mathfrak{R}[a, b]$ *and*

$$\left| \int_a^b f(x)dx \right| \leq \int_a^b |f(x)|dx.$$

Proof. (1) Since $f, g \in \mathfrak{R}[a, b]$, for every $\varepsilon > 0$, there exist $P_1, P_2 \in \wp[a, b]$ such that

$$U(f, P_1) - L(f, P_1) < \frac{\varepsilon}{2}, \quad U(g, P_2) - L(g, P_2) < \frac{\varepsilon}{2}.$$

Put $P = P_1 \cup P_2$. Then, we have $P \in \wp[a, b]$ and

$$
\begin{aligned}
U(f + g, P) - L(f + g, P) &\leq U(f, P) + U(g, P) - L(f, P) - L(g, P) \\
&= U(f, P) - L(f, P) + U(g, P) - L(g, P) \\
&\leq U(f, P_1) - L(f, P_1) + U(g, P_2) - L(g, P_2) \\
&< \frac{\varepsilon}{2} + \frac{\varepsilon}{2} \\
&= \varepsilon.
\end{aligned}
$$

Therefore, by the Riemann Integral Test, we have $f + g \in \mathfrak{R}[a, b]$.

On the other hand, it follows that, for every $P \in \wp[a, b]$,

$$
\begin{aligned}
L(f, P) + L(g, P) &\leq L(f + g, P) \\
&\leq \int_a^b (f + g)(x)dx \\
&\leq U(f + g, P) \\
&\leq U(f, P) + U(g, P).
\end{aligned}
\tag{8.2.3}
$$

Observe that

$$U(f,P) \le U(f,P_1) < L(f,P_1) + \frac{\varepsilon}{2} \le \int_a^b f(x)dx + \frac{\varepsilon}{2}$$

and

$$U(g,P) \le U(g,P_2) < L(g,P_2) + \frac{\varepsilon}{2} \le \int_a^b g(x)dx + \frac{\varepsilon}{2}.$$

Thus, by (8.2.3), we have

$$\int_a^b (f+g)(x)dx \le U(f+g,P)$$
$$\le U(f,P) + U(g,P)$$
$$< \int_a^b f(x)dx + \int_a^b g(x)dx + \varepsilon.$$

Since $\varepsilon > 0$ is arbitrary, we have

$$\int_a^b (f+g)(x)dx \le \int_a^b f(x)dx + \int_a^b g(x)dx. \tag{8.2.4}$$

Similarly, we have

$$\int_a^b (f+g)(x)dx \ge \int_a^b f(x)dx + \int_a^b g(x)dx. \tag{8.2.5}$$

Therefore, from (8.2.4) and (8.2.5), it follows that

$$\int_a^b (f+g)(x)dx = \int_a^b f(x)dx + \int_a^b g(x)dx.$$

(2) It is easy to prove (2) by the definition of the Riemann integral.

(3) Let $f \in \Re[a,b]$ and $h(x) = x^2$ for all $x \in [a,b]$. Then, we have

$$h \circ f = f^2 \in \Re[a,b].$$

Next, let $f, g \in \Re[a,b]$. Then, by (1) and (2), we have

$$-g, \ f+g, \ f-g, \ (f+g)^2, \ (f-g)^2 \in \Re[a,b].$$

Therefore, since

$$fg = \frac{1}{4}[(f+g)^2 - (f-g)^2],$$

we have $fg \in \Re[a,b]$.

(4) Since $g(x) = |x|$ is continuous on $[a,b]$, we have $|f| = g \circ f \in \Re[a,b]$. On the other hand, let $c \in \mathbb{R}$ be such that

$$c \int_a^b f(x)dx = \left| \int_a^b f(x)dx \right|.$$

Then, since $cf(x) \leq |f(x)|$ for every $x \in [a, b]$, we have

$$\int_a^b cf(x)dx \leq \int_a^b |f(x)|dx.$$

Therefore, by (2), we have

$$\left| \int_a^b f(x)dx \right| = c \int_a^b f(x)dx = \int_a^b cf(x)dx \leq \int_a^b |f(x)|dx.$$

This completes the proof.

From Theorem 8.2.2, we know that

$$\Re[a, b] = \{f : [a, b] \to \mathbb{R} : \text{ a Riemann integrable function on } [a, b]\}$$

is a vector space.

Theorem 8.2.3. *Let $f : [a, b] \to \mathbb{R}$ be a bounded function on $[a, b]$, and let a number c with $a < c < b$. Then, the following are equivalent:*

(1) *f is Riemann integrable on $[a, b]$;*

(2) *f is Riemann integrable on both $[a, c]$ and $[c, b]$. In this case, we have*

$$\int_a^b f(x)dx = \int_a^c f(x)dx + \int_c^b f(x)dx.$$

Proof. (2) $\Longrightarrow$ (1) Suppose that f is Riemann integrable on both $[a, c]$ and $[c, b]$. Then, by the Riemann Integral Test, for every $\varepsilon > 0$, there exist $P(\varepsilon)_1 \in \wp[a, c]$ and $P(\varepsilon)_2 \in \wp[c, b]$ such that

$$U(f, P(\varepsilon)_1) - L(f, P(\varepsilon)_1) < \frac{\varepsilon}{2}$$

and

$$U(f, P(\varepsilon)_2) - L(f, P(\varepsilon)_2) < \frac{\varepsilon}{2}.$$

Let $P(\varepsilon) = P(\varepsilon)_1 \cup P(\varepsilon)_2$. Then, we have

$$\begin{aligned}
&U(f, P(\varepsilon)) - L(f, P(\varepsilon)) \\
&= [U(f, P(\varepsilon)_1) + U(f, P(\varepsilon)_2)] - [L(f, P(\varepsilon)_1) + L(f, P(\varepsilon)_2)] \\
&= [U(f, P(\varepsilon)_1) - L(f, P(\varepsilon)_1)] + [U(f, P(\varepsilon)_2) - L(f, P(\varepsilon)_2)] \\
&< \frac{\varepsilon}{2} + \frac{\varepsilon}{2} = \varepsilon.
\end{aligned}$$

Since $\varepsilon > 0$ is arbitrary, by the Riemann Integral Test, f is Riemann integrable on $[a, b]$.

(1) $\Longrightarrow$ (2) Suppose that f is Riemann integrable on $[a, b]$. Then, by the Riemann Integral Test, for every $\varepsilon > 0$, there exists $P_1 \in \wp[a, b]$ such that

$$U(f, P_1) - L(f, P_1) < \varepsilon.$$

If $P = P_1 \cup \{c\}$, then P is a refinement of P_1 and so

$$U(f, P) - L(f, P) \le U(f, P_1) - L(f, P_1) < \varepsilon. \tag{8.2.6}$$

Let $P_2 = P \cap [a, c]$ and $P_3 = P \cap [c, b]$. Then, we have

$$U(f, P) = U(f, P_2) + U(f, P_3), \quad L(f, P) = L(f, P_2) + L(f, P_3)$$

and so, from (8.2.6), it follows that

$$[U(f, P_2) - L(f, P_2)] + [U(f, P_3) - L(f, P_3)] < \varepsilon.$$

Thus, we have

$$U(f, P_2) - L(f, P_2) < \varepsilon, \quad U(f, P_3) - L(f, P_3) < \varepsilon. \tag{8.2.7}$$

Since $\varepsilon > 0$ is arbitrary, by the Riemann Integral Test, f is Riemann integrable on both $[a, c]$ and $[c, b]$.

Finally, we prove

$$\int_a^b f(x)dx = \int_a^c f(x)dx + \int_c^b f(x)dx.$$

In fact, from (8.2.7), it follows that

$$\int_a^b f(x)dx \le U(f, P) = U(f, P_2) + U(f, P_3)$$
$$< L(f, P_2) + L(f, P_3) + 2\varepsilon \tag{8.2.8}$$
$$\le \int_a^c f(x)dx + \int_c^b f(x)dx + 2\varepsilon.$$

Also, from from (8.2.7), it follows that

$$\int_a^c f(x)dx + \int_c^b f(x)dx \le U(f, P_2) + U(f, P_3)$$
$$< L(f, P_2) + L(f, P_3) + 2\varepsilon$$
$$= L(f, P) + 2\varepsilon \tag{8.2.9}$$
$$\le \int_a^b f(x)dx + 2\varepsilon.$$

Since $\varepsilon > 0$ is arbitrary, it follows from (8.2.8) and (8.2.9) that

$$\int_a^b f(x)dx = \int_a^c f(x)dx + \int_c^b f(x)dx.$$

This completes the proof.

8.3   The Fundamental Theorems of Calculus

In this section, we prove the Fundamental Theorem of Calculus for a correction between the derivative and the Riemann integral.

Theorem 8.3.1. *Let $f : [a, b] \to \mathbb{R}$ be Riemann integrable on $[a, b]$, and define a function $F : [a, b] \to \mathbb{R}$ by*

$$F(x) = \int_a^x f(t)dt \quad \text{for every} \ \ x \in [a, b].$$

Then, we have the following:

(1) *F is uniformly continuous on $[a, b]$;*

(2) *If f is continuous at $x_0 \in [a, b]$, then F is differentiable at x_0 and $F'(x_0) = f(x_0)$.*

Proof. (1) Since f is bounded on $[a, b]$, there exists a constant $M > 0$ such that

$$|f(t)| \le M \quad \text{for every} \ \ t \in [a, b].$$

Further, for every $x, y \in [a, b]$, we have

$$|F(x) - F(y)| = \left| \int_x^y f(t)dt \right| \le \int_x^y |f(t)|dt \le M|y - x|.$$

Therefore, for every $\varepsilon > 0$, if $\delta = \frac{\varepsilon}{M}$, then

$$|y - x| < \delta, \ \ x, y \in [a, b] \implies |F(y) - F(x)| < \varepsilon,$$

which implies that F is uniformly continuous on $[a, b]$.

(2) Let $\varepsilon > 0$. Since f is continuous at $x = x_0$, there exists $\delta > 0$ such that

$$|x - x_0| < \delta, \ \ x \in [a, b] \implies |f(x) - f(x_0)| < \varepsilon.$$

Thus, if $0 < |x - x_0| < \delta$, then we have

$$\left| \frac{F(x) - F(x_0)}{x - x_0} - f(x_0) \right| = \left| \frac{1}{x - x_0} \left(\int_0^x f(t)dt - \int_0^{x_0} f(t)dt \right) - f(x_0) \right|$$

$$\le \frac{1}{x - x_0} \int_{x_0}^x |f(t) - f(x_0)|dt < \varepsilon.$$

Therefore, F is differentiable at x_0 and $F'(x_0) = f(x_0)$. This completes the proof.

Theorem 8.3.2. (The Fundamental Theorem of Calculus I) *Let f : $[a, b] \to \mathbb{R}$ be a continuous function on $[a, b]$, and define a function $F : [a, b] \to \mathbb{R}$ by*

$$F(x) = \int_a^x f(t)dt \quad \text{for every} \ \ x \in [a, b].$$

Then, F is differentiable on $[a, b]$ and $F'(x) = f(x)$.

Proof. Since f is continuous on $[a, b]$, for every $\varepsilon > 0$, there exists $\delta > 0$ such that

$$|t - x| < \delta \implies |f(t) - f(x)| < \varepsilon.$$

Let h be a number with $0 < h < \delta$. Then, we have

$$F(x + h) - F(x) = \int_a^{x+h} f(t)dt - \int_a^x f(t)dt = \int_x^{x+h} f(t)dt$$

and so

$$\left| \frac{F(x + h) - F(x)}{h} - f(x) \right| = \left| \frac{1}{h} \int_x^{x+h} f(t)dt - \frac{1}{h} \int_x^{x+h} f(x)dt \right|$$

$$\leq \frac{1}{h} \int_x^{x+h} |f(t) - f(x)|dt < \varepsilon.$$

Therefore, we have

$$F'_+(x) = \lim_{h \to 0^+} \frac{F(x + h) - F(x)}{h} = f(x).$$

Similarly, for $-\delta < h < 0$, we have

$$F'_-(x) = \lim_{h \to 0^-} \frac{F(x + h) - F(x)}{h} = f(x).$$

Since $F'_+(x) = F'_-(x) = f(x)$, F is differentiable on $[a, b]$ and, for all $x \in [a, b]$, $F'(x) = f(x)$. This completes the proof.

Theorem 8.3.3. (The Fundamental Theorem of Calculus II) *Let f : $[a, b] \to \mathbb{R}$ be Riemann integrable on $[a, b]$ and F be differentiable on $[a, b]$. If $F'(x) = f(x)$, then we have*

$$\int_a^b f(x)dx = [F(x)]_a^b = F(b) - F(a).$$

Proof. Let $P = \{a = x_0, x_1, x_2, \ldots, x_n = b\}$ be a partition of $[a, b]$. Then, if we apply Lagrange's Mean Value Theorem to F at each subinterval $I_i = [x_{i-1}, x_i]$ for every $i = 1, 2, \ldots, n$, then it follows that there exists $t_i \in [x_{i-1}, x_i]$ such that

$$F(x_i) - F(x_{i-1}) = f(t_i)(x_i - x_{i-1}).$$

Thus, we have

$$F(b) - F(a) = \sum_{i=1}^{n}(F(x_i) - F(x_{i-1})) = \sum_{i=1}^{n} f(t_i)(x_i - x_{i-1}).$$

Put

$$M_i = \sup\{f(x) : x \in [x_{i-1}, x_i]\}, \quad m_i = \inf\{f(x) : x \in [x_{i-1}, x_i]\}.$$

Then, we have $m_i \le f(t_i) \le M_i$ for every $i = 1, 2, \ldots, n$ and so

$$\sum_{i=1}^{n} m_i(x_i - x_{i-1}) \le F(b) - F(a) \le \sum_{i=1}^{n} M_i(x_i - x_{i-1}).$$

Thus, it follows that, for every $P \in \wp[a, b]$,

$$L(f, P) \le F(b) - F(a) \le U(f, P)$$

and so, since

$$\sup\{L(f, P) : P \in \wp[a, b]\} \le F(b) - F(a) \le \inf\{U(f, P) : P \in \wp[a, b]\}$$

and $f \in \Re[a, b]$, we have

$$\int_{a}^{b} f(x)dx = F(b) - F(a).$$

This completes the proof.

Example 8.3.1. Find an example in which if F is not differentiable on $[a, b]$, then Theorem 8.3.3 is not true.

Solution. Define two functions $f, F : [0, 1] \to \mathbb{R}$ by $f(x) = 1$ for every $x \in [0, 1]$ and

$$F(x) = \begin{cases} x, & \text{if } x \in [0, 1), \\ 0, & \text{if } x = 1, \end{cases}$$

respectively. Then, f is Riemann integrable on $[0, 1]$ and $\int_{0}^{1} f(x)dx = 1$.

On the other hand, we have $F(1) = F(0) = 0$, and, for every $x \in [0, 1)$, $F'(x) = f(x)$, but F is not differentiable at $x = 1$. Thus, Theorem 8.3.3 is not true.

Theorem 8.3.4. (The Generalized Mean Value Theorem for Integral)
Let $f : [a, b] \to \mathbb{R}$ be a continuous function on $[a, b]$ and $g : [a, b] \to \mathbb{R}$ be Riemann integrable on $[a, b]$ with $g(x) \ge 0$ for every $x \in [a, b]$. Then, there exists $c \in [a, b]$ such that

$$\int_{a}^{b} f(x)g(x)dx = f(c) \int_{a}^{b} g(x)dx.$$

Proof. Since $f : [a, b] \to \mathbb{R}$ is continuous on $[a, b]$, let

$$M = \sup\{f(x) : x \in [a, b]\}, \quad m = \inf\{f(x) : x \in [a, b]\}.$$

Then, it follows that

$$mg(x) \le f(x)g(x) \le Mg(x) \quad \text{for every} \quad x \in [a, b]$$

and fg is Riemann integrable on $[a, b]$ and so

$$m \int_a^b g(x)dx \le \int_a^b f(x)g(x)dx \le M \int_a^b g(x)dx. \qquad (8.3.1)$$

If $\int_a^b g(x)dx = 0$, then the conclusion follows easily.

If $\int_a^b g(x)dx \ne 0$, then it follows from (8.3.1) that

$$m \le \frac{\int_a^b f(x)g(x)dx}{\int_a^b g(x)dx} \le M$$

and so, since f is continuous on $[a, b]$, by the Intermediate Value Theorem, there exists $c \in [a, b]$ such that

$$f(c) = \frac{\int_a^b f(x)g(x)dx}{\int_a^b g(x)dx},$$

that is,

$$\int_a^b f(x)g(x)dx = f(c) \int_a^b g(x)dx.$$

This completes the proof.

Example 8.3.2. Define two functions $f, g : [-1, 1] \to \mathbb{R}$ by

$$f(x) = x, \quad g(x) = e^x \quad \text{for every} \quad x \in [-1, 1].$$

Then, apply Theorem 8.3.4 to the functions f and g on $[-1, 1]$.

Solution. Since f and g are continuous on $[-1, 1]$, they are Riemann integrable on $[-1, 1]$. Further, $g(x) > 0$ for every $x \in [-1, 1]$ and

$$\int_{-1}^1 f(x)g(x)dx = \frac{2}{e}, \quad \int_{-1}^1 g(x)dx = \frac{e^2 - 1}{e}$$

and so

$$\frac{2}{e^2 - 1} = \frac{\int_{-1}^1 f(x)g(x)dx}{\int_{-1}^1 g(x)dx}.$$

Since f is continuous on $[-1, 1]$ and

$$f(-1) = -1 < \frac{2}{e^2 - 1} < 1 = f(1),$$

by the Intermediate Value Theorem, there exists $c \in [-1, 1]$ such that

$$c = f(c) = \frac{2}{e^2 - 1}.$$

Therefore, we have

$$f(c) \int_{-1}^{1} g(x)dx = \frac{2}{e^2 - 1} \cdot \frac{e^2 - 1}{e} = \frac{2}{e} = \int_{-1}^{1} f(x)g(x)dx.$$

From Theorem 8.3.4, we have the following result:

Corollary 8.3.5. (The Mean Value Theorem for Integral) *Let* $f :$ *$[a, b] \to \mathbb{R}$ be a continuous function on $[a, b]$. Then, there exists $c \in [a, b]$ such that*

$$\int_{a}^{b} f(x)dx = f(c)(b - a).$$

Proof. In Theorem 8.3.4, if $g(x) = 1$ for every $x \in [a, b]$, then we have

$$\int_{a}^{b} f(x)dx = f(c)(b - a).$$

Example 8.3.3. By using the Mean Value Theorem for Integral, prove the following inequalities:

$$\frac{x}{1 + x} < \ln(1 + x) < x \quad \text{for every} \quad x > 0.$$

Solution. Let $f(x) = \frac{1}{1+x}$ for every $x > 0$. Then, we have

$$\int_{0}^{x} \frac{1}{1 + t}dt = \ln(1 + x).$$

Thus, by Corollary 8.3.5, there exists $c \in [0, x]$ such that

$$\int_{0}^{x} \frac{1}{1 + t}dt = f(c)(x - 0).$$

Since f is decreasing on $[0, x]$ and

$$\inf\{f(t) : t \in [0, x]\} \leq f(c) \leq \sup\{f(t) : t \in [0, x]\},$$

we have

$$\frac{1}{1 + x} < \frac{\ln(1 + x)}{x} < 1 \quad \text{for every} \quad x > 0,$$

which implies that

$$\frac{x}{1 + x} < \ln(1 + x) < x \quad \text{for every} \quad x > 0.$$

8.4   The Substitution Theorem and Integration by Parts

In this section, we prove the Substitution Theorem and the Integration by Parts as techniques of integration, which are based on the Fundamental Theorems of Calculus.

Theorem 8.4.1. (The Substitution Theorem) *Let* $f : [a, b] \to \mathbb{R}$ *be a continuous function on* $[a, b]$ *and* $g : [c, d] \to [a, b]$ *be differentiable on* $[c, d]$. *If* $g(c) = a$ *and* $g(d) = b$, *then we have*

$$\int_a^b f(x)dx = \int_c^d f(g(t))g'(t)dt.$$

Proof. For every $x \in [a, b]$, define a function $F : [a, b] \to \mathbb{R}$ by

$$F(x) = \int_a^x f(t)dt.$$

Then, by Theorem 8.3.2 (The Fundamental Theorem of Calculus I), F is differentiable on $[a, b]$. Thus, we have

$$(F(g(t)))' = F'(g(t))g'(t) = f(g(t))g'(t).$$

Therefore, by Theorem 8.3.3 (The Fundamental Theorem of Calculus II), we have

$$\int_c^d f(g(t))g'(t)dt = F(g(d)) - F(g(c)) = F(b) - F(a) = \int_a^b f(x)dx.$$

This completes the proof.

Example 8.4.1. Let $f : [0, 1] \to \mathbb{R}$ be a continuous function on $[0, 1]$. Evaluate the following:

$$\int_{-1}^1 xf(x^2)dx.$$

Solution. Let $\phi(x) = x^2$ for every $x \in [-1, 1]$. Then, since we have $\phi([-1, 1]) = [0, 1]$ and f is continuous on $[0, 1]$, we have

$$\int_{-1}^1 f(x^2)dx = \frac{1}{2}\int_{-1}^1 f(\phi(x))\phi'(x)dx = \frac{1}{2}\int_1^1 f(t)dt = 0.$$

Theorem 8.4.2. (The Integration by Parts) *Let* $f, g : [a, b] \to \mathbb{R}$ *be differentiable on* $[a, b]$ *and* f', g' *be Riemann integrable on* $[a, b]$. *Then, we have*

$$\int_a^b f(x)g'(x)dx = f(b)g(b) - f(a)g(a) - \int_a^b f'(x)g(x)dx.$$

Proof. Since f and g are differentiable on $[a, b]$, fg is also differentiable on $[a, b]$ and so, for every $x \in [a, b]$,

$$(f(x)g(x))' = f'(x)g(x) + f(x)g'(x).$$

Since f and g are continuous on $[a, b]$, they are Riemann integrable on $[a, b]$. Since f' and g' are Riemann integrable on $[a, b]$, $(fg)' = f'g + fg'$ is also Riemann integrable on $[a, b]$. Therefore, by Theorem 8.3.2 (The Fundamental Theorem of Calculus I), we have

$$\int_a^b (f(x)g(x))' dx = f(b)g(b) - f(a)g(a)$$

and so

$$\int_a^b f'(x)g(x)dx + \int_a^b f(x)g'(x)dx = f(b)g(b) - f(a)g(a),$$

which implies that

$$\int_a^b f(x)g'(x)dx = f(b)g(b) - f(a)g(a) - \int_a^b f'(x)g(x)dx.$$

This completes the proof.

Example 8.4.2. Evaluate the following:

$$\int_0^{\frac{\pi}{2}} \cos^3 x dx.$$

Solution. Now, we have

$$\int_0^{\frac{\pi}{2}} \cos^3 x dx = \int_0^{\frac{\pi}{2}} \cos^2 x (\cos x dx)$$

$$= \int_0^{\frac{\pi}{2}} (1 - \sin^2 x)(\cos x dx).$$

Let $u = \sin x$. Then, we have $du = \cos x dx$. If $x = 0$, then $u = 0$. If $x = \frac{\pi}{2}$, then $u = 1$. Therefore, we have

$$\int_0^{\frac{\pi}{2}} \cos^3 x dx = \int_0^1 (1 - u^2) du$$

$$= \left[u - \frac{1}{3} u^3 \right]_0^1$$

$$= 1 - \frac{1}{3} = \frac{2}{3}.$$

8.5 Improper Integrals

In the previous sections, all of the functions have been bounded and all of integrals have been computed on closed and bounded intervals. In this section, we relax these restrictions by defining improper Riemann integrals.

If $f : [a, b] \to \mathbb{R}$ be a Riemann integrable function on $[a, b]$, that is, $f \in \wp[a, b]$, then we have

$$\int_a^b f(x)dx = \lim_{c \to a^+} \left(\lim_{d \to b^-} \int_c^d f(x)dx \right) = \lim_{d \to b^-} \left(\lim_{c \to a^+} \int_c^d f(x)dx \right).$$

In fact, define a function $F : [a, b] \to \mathbb{R}$ by

$$F(x) = \int_a^x f(t)dt \text{ for every } x \in [a, b].$$

Then, F is continuous on $[a, b]$ and so

$$\int_a^b f(x)dx = F(b) - F(a)$$

$$= \lim_{d \to b^-} F(d) - \lim_{c \to a^+} F(c)$$

$$= \lim_{c \to a^+} \left(\lim_{d \to b^-} (F(d) - F(c)) \right)$$

$$= \lim_{c \to a^+} \left(\lim_{d \to b^-} \int_c^d f(x)dx \right).$$

Similarly, we have

$$\int_a^b f(x)dx = \lim_{d \to b^-} \left(\lim_{c \to a^+} \int_c^d f(x)dx \right).$$

But, in general, the converse is not true. If the converse is true, then f is said to be *improper Riemann integrable* on $[a, b]$. This means that if $f : [a, b] \to \mathbb{R}$ is Riemann integrable on $[a, b]$, then f is improper Riemann integrable on $[a, b]$, and so the improper Riemann integral is an extension of the Riemann integral.

Definition 8.5.1. (1) Let $a, b \in \mathbb{R} = (-\infty, +\infty)$ with $a < b$ and $f : (a, b] \to \mathbb{R}$ be a function. If $f \in \Re[c, b]$ for every $c \in (a, b)$, then the *improper Riemann integral* of f on $[a, b]$ is defined by

$$\int_a^b f(x)dx = \lim_{c \to a^+} \int_c^b f(x)dx$$

provided the limit exists. Then, we say that $\int_a^b f(x)dx$ is *convergent*. Otherwise, we say that $\int_a^b f(x)dx$ is *divergent*;

(2) Let $a, b \in \mathbb{R} = (-\infty, +\infty)$ with $a < b$ and $f : [a, b) \to \mathbb{R}$ be a function. If $f \in \Re[a, c]$ for every $c \in (a, b)$, then the *improper Riemann integral* of f on $[a, b]$ is defined by

$$\int_a^b f(x)dx = \lim_{c \to b^-} \int_a^c f(x)dx$$

provided the limit exists. Then, we say that $\int_a^b f(x)dx$ is *convergent*. Otherwise, we say that $\int_a^b f(x)dx$ is *divergent*;

(3) Let $a, b \in \mathbb{R} = (-\infty, +\infty)$ with $a < b$ and $f : (a, b) \to \mathbb{R}$ be a function. If $f \in \Re[c, d]$ for every $c, d \in (a, b)$ with $c < d$, then the *improper Riemann integral* of f on $[a, b]$ is defined by

$$\int_a^b f(x)dx = \lim_{c \to a^+} \left(\lim_{d \to b^-} \int_c^d f(x)dx \right) = \lim_{d \to b^-} \left(\lim_{c \to a^+} \int_c^d f(x)dx \right)$$

provided the limits exist. Then, we say that $\int_a^b f(x)dx$ is *convergent*. Otherwise, we say that $\int_a^b f(x)dx$ is *divergent*;

(4) If $c \in (a, b)$ be such that $f \in \Re[a, d]$ and $f \in \Re[e, b]$ for every $d \in (a, c)$ and $e \in (c, b)$, then the *improper Riemann integral* of f on $[a, b]$ is defined by

$$\int_a^b f(x)dx = \lim_{d \to c^-} \int_a^d f(x)dx + \lim_{e \to c^+} \int_e^b f(x)dx$$

provided the limits exist, that is, the improper Riemann integrals

$$\int_a^c f(x)dx, \quad \int_c^b f(x)dx$$

are convergent.

Example 8.5.1. Let $f : (0, 2] \to \mathbb{R}$ be a function defined by $f(x) = \frac{1}{\sqrt{x}}$ for every $x \in (0, 2]$. Show that f is an improper Riemann integrable function on $[0, 2]$.

Solution. For every $c \in (0, 2]$, since f is continuous on $(0, 2]$, f is Riemann integrable on $[0, 2]$. Further, we have

$$\int_0^2 \frac{1}{\sqrt{x}}dx = \lim_{c \to 0^+} \int_c^2 \frac{1}{\sqrt{x}}dx = \lim_{c \to 0^+} (2\sqrt{2} - 2\sqrt{c}) = 2\sqrt{2}$$

and so f is an improper Riemann integrable function on $[0, 2]$.

Example 8.5.2. Let $f : (0, 1] \to \mathbb{R}$ be a function defined by $f(x) = \frac{1}{x}$ for every $x \in (0, 1]$. Show that f is not an improper Riemann integrable function on $[0, 1]$.

Solution. The function f is continuous and Riemann integrable on $(0, 1]$, but $f(x)$ is not defined at $x = 0$. For any $c \in (0, 1]$, we have

$$\int_0^1 \frac{1}{x} dx = [\ln x]_c^1 = \ln 1 - \ln c = -\ln c,$$

but the limit

$$\lim_{c \to 0^+} \int_c^1 \frac{1}{x} dx = \lim_{c \to 0^+} (-\ln c) = \infty$$

does not exist. Therefore, $f(x) = \frac{1}{x}$ is not an improper Riemann integrable function on $[0, 1]$.

Definition 8.5.2. (1) Let $f : [a, \infty) \to \mathbb{R}$ be a function. If $f \in \Re[a, b]$ for every $a < b$, then the *improper Riemann integral* of f on $[a, \infty)$ is defined by

$$\int_a^\infty f(x) dx = \lim_{b \to \infty} \int_a^b f(x) dx$$

provided the limit exists. Then, we say that $\int_a^\infty f(x) dx$ is *convergent*. Otherwise, we say that $\int_a^\infty f(x) dx$ is *divergent*;

(2) If $f \in \Re[a, b]$ for every $a < b$, then the *improper Riemann integral* of f on $(-\infty, b]$ is defined by

$$\int_{-\infty}^b f(x) dx = \lim_{a \to -\infty} \int_a^b f(x) dx$$

provided the limit exists. Then, we say that $\int_{-\infty}^b f(x) dx$ is *convergent*. Otherwise, we say that $\int_{-\infty}^b f(x) dx$ is *divergent*;

(3) If $f : \mathbb{R} \to \mathbb{R}$ is a function and $f \in \Re[a, b]$ for every $a, b \in \mathbb{R} = (-\infty, \infty)$ with $a < b$, then the *improper Riemann integral* of f on $(-\infty, \infty)$ is defined by

$$\int_{-\infty}^\infty f(x) dx = \lim_{a \to -\infty} \int_a^c f(x) dx + \lim_{b \to \infty} \int_c^b f(x) dx$$

for every $c \in \mathbb{R}$ provided the limits exist, that is, the improper Riemann integrals

$$\int_c^\infty f(x) dx, \quad \int_{-\infty}^c f(x) dx$$

are convergent.

Example 8.5.3. Evaluate the following:

$$\int_1^\infty \frac{1}{\sqrt{x}} dx.$$

Solution. By the definition of the improper Riemann integral, we have

$$\int_1^\infty \frac{1}{\sqrt{x}} dx = \lim_{b\to\infty} \int_1^b \frac{1}{\sqrt{x}} dx = \lim_{b\to\infty} 2[\sqrt{b}-1] = \infty.$$

Thus, $\int_1^\infty \frac{1}{\sqrt{x}} dx$ is divergent.

Example 8.5.4. Evaluate the following:

$$\int_{-\infty}^2 \frac{1}{(4-x)^2} dx.$$

Solution. By the definition of the improper Riemann integral, we have

$$\int_{-\infty}^2 \frac{1}{(4-x)^2} dx = \lim_{a\to-\infty} \int_a^2 \frac{1}{(4-x)^2} dx$$

$$= \lim_{a\to-\infty} \left[\frac{1}{4-x}\right]_a^2$$

$$= \lim_{a\to-\infty} \left(\frac{1}{2} - \frac{1}{4-a}\right)$$

$$= \frac{1}{2} - 0 = \frac{1}{2}.$$

Thus, $\int_{-\infty}^2 \frac{1}{(4-x)^2} dx$ is convergent.

Example 8.5.5. Evaluate the following:

$$\int_{-\infty}^\infty xe^{-x^2} dx.$$

Solution. By the definition of the improper Riemann integral, we have

$$\int_{-\infty}^\infty xe^{-x^2} dx = \int_{-\infty}^0 xe^{-x^2} dx + \int_0^\infty xe^{-x^2} dx$$

$$= \lim_{a\to-\infty} \int_a^0 xe^{-x^2} dx + \lim_{b\to\infty} \int_0^b xe^{-x^2} dx$$

$$= \lim_{a\to-\infty} \left(-\frac{1}{2} + \frac{1}{2}e^{-a^2}\right) + \lim_{b\to\infty} \left(-\frac{1}{2}e^{-b^2} + \frac{1}{2}\right)$$

$$= -\frac{1}{2} + \frac{1}{2} = 0.$$

Thus, $\int_{-\infty}^\infty xe^{-x^2} dx$ is convergent.

Theorem 8.5.1. *If the improper Riemann integrals*

$$\int_a^\infty f(x)dx, \quad \int_a^\infty g(x)dx$$

are convergent, then, for every $\alpha, \beta \in \mathbb{R}$, $\int_a^\infty (\alpha f(x) + \beta g(x))dx$ is convergent.

Proof. Note that the limit

$$\lim_{b \to \infty} \int_a^b (\alpha f(x) + \beta g(x))dx = \lim_{b \to \infty} \left(\int_a^b \alpha f(x)dx + \int_a^b \beta g(x)dx \right)$$

$$= \alpha \lim_{b \to \infty} \int_a^b f(x)dx + \beta \lim_{b \to \infty} \int_a^b g(x)dx$$

exists. Therefore, $\int_a^\infty (\alpha f(x) + \beta g(x))dx$ is convergent and

$$\int_a^\infty (\alpha f(x) + \beta g(x))dx = \alpha \int_a^\infty f(x)dx + \beta \int_a^\infty g(x)dx.$$

This completes the proof.

Theorem 8.5.2. *Let* $f : [a, \infty) \to \mathbb{R}$ *be a function such that* $f(x) \geq 0$ *for every* $x \in [a, \infty)$ *and* f *is Riemann integrable on* $[a, b]$*. Then, the following are equivalent:*

(1) *The improper Riemann integral* $\int_a^\infty f(x)dx$ *is convergent;*

(2) *The set*

$$\left\{ \int_a^b f(x)dx : b \in (a, \infty) \right\}$$

is bounded and, further,

$$\int_a^\infty f(x)dx = \sup \left\{ \int_a^b f(x)dx : b \in (a, \infty) \right\}.$$

Proof. (1) $\implies$ (2) Define a function $F : [a, \infty) \to \mathbb{R}$ by

$$F(b) = \int_a^b f(x)dx,$$

and suppose that the improper Riemann integral $\int_a^\infty f(x)dx$ is convergent. Since F is increasing, for every $b \in [a, \infty)$, we have

$$F(b) = \int_a^b f(x)dx \leq \int_a^\infty f(x)dx$$

and so

$$\left\{ \int_a^b f(x)dx : b \in (a, \infty) \right\}$$

is bounded.

(2) $\implies$ (1) Suppose that

$$\left\{ \int_a^b f(x)dx : b \in (a, \infty) \right\}$$

is bounded. Let $L = \sup\{F(b) : b \in (a, \infty)\}$. Then, for every $\varepsilon > 0$, there exists a number $b \in \mathbb{R}$ such that $L - \varepsilon < F(b)$. Let $x \geq b$. Since $f(x) \geq 0$, we have

$$L - \varepsilon < F(b) \leq F(x) \leq L < L + \varepsilon.$$

Since F is increasing and bounded from above, we have

$$\lim_{b \to \infty} F(b) = \sup\{F(b) : b \in (a, \infty)\}$$

and

$$\int_a^\infty f(x)dx = \sup\left\{\int_a^b f(x)dx : b \in (a, \infty)\right\}.$$

This completes the proof.

Theorem 8.5.2. (The Comparison Test for Integral) *Let $f, g : [a, \infty) \to \mathbb{R}$ be two functions such that, for every $x \in [a, \infty)$, $0 \leq f(x) \leq g(x)$ and $\int_a^\infty g(x)dx$ is convergent. Then, $\int_a^\infty f(x)dx$ is convergent and*

$$\int_a^\infty f(x)dx \leq \int_a^\infty g(x)dx.$$

Proof. Let $b \in [a, \infty)$. Then, we have

$$\int_a^b f(x)dx \leq \int_a^b g(x)dx \leq \int_a^\infty g(x)dx.$$

Since $\int_a^\infty g(x)dx$ is convergent, by Theorem 8.5.1, it follows that $\int_a^\infty f(x)dx$ is convergent and

$$\int_a^\infty f(x)dx \leq \int_a^\infty g(x)dx.$$

This completes the proof.

Example 8.5.6. Show that $\int_1^\infty \sqrt{x}e^{-x^2}dx$ is convergent.

Solution. For every $x \geq 1$, we have

$$0 \leq \sqrt{x}e^{-x^2} \leq xe^{-x^2}$$

and

$$\int_1^\infty xe^{-x^2}dx = \lim_{b \to \infty} \int_1^b xe^{-x^2}dx = \lim_{b \to \infty}\left(-\frac{1}{2}e^{-b^2} + \frac{1}{2e}\right) = \frac{1}{2e}.$$

Therefore, by Theorem 8.5.2, it follows that $\int_1^\infty \sqrt{x}e^{-x^2}dx$ is convergent.

Theorem 8.5.3. *Let $f : [a, \infty) \to \mathbb{R}$ be a function such that, for every $b \in [a, \infty)$, f is Riemann integrable on $[a, b]$ and $\int_a^\infty |f(x)|dx$ is convergent. Then, $\int_a^\infty f(x)dx$ is convergent and*

$$\left| \int_a^\infty f(x)dx \right| \leq \int_a^\infty |f(x)|dx.$$

Proof. For every $x \in [a, \infty)$, we have

$$-|f(x)| \leq f(x) \leq |f(x)|$$

and so

$$0 \leq f(x) + |f(x)| \leq 2|f(x)|.$$

Thus, for every $b \in [a, \infty)$, $f + |f|$ is Riemann integrable on $[a, b]$. Since $\int_a^\infty 2|f(x)|dx$ is convergent, by Theorem 8.5.2, $\int_a^\infty (f(x) + |f(x)|)dx$ is convergent. On the other hand, for every $x \in [a, \infty)$, since we have

$$f(x) = f(x) + |f(x)| - |f(x)|,$$

it follows that $\int_a^\infty f(x)dx$ is convergent.

Now, let $\int_a^\infty |f(x)|dx = L$. Then, since $-|f(x)| \leq f(x) \leq |f(x)|$ for every $x \in [a, \infty)$, we have

$$-L = -\int_a^\infty |f(x)|dx \leq \int_a^\infty f(x)dx \leq \int_a^\infty |f(x)|dx = L$$

and so

$$\left| \int_a^\infty f(x)dx \right| \leq L = \int_a^\infty |f(x)|dx.$$

This completes the proof.

Definition 8.5.3. (1) An improper integral $\int_a^b f(x)dx$ is said to be *absolutely convergent* if the improper integral $\int_a^b |f(x)|dx$ converges;

(2) An improper integral $\int_a^b f(x)dx$ is said to be *conditionally convergent* if $\int_a^b f(x)dx$ converges, but $\int_a^b |f(x)|dx$ diverges.

8.6 The Riemann–Stieltjes Integral

In this section, we give an extension of the Riemann integral, which is called the Riemann–Stieltjes integral, and discuss some properties of the Riemann–Stieltjes integral.

Definition 8.6.1. Let $P = \{a = x_0, x_1, x_2, \ldots, x_n = b\}$ be a partition of $[a, b]$ and $f : [a, b] \to \mathbb{R}$ be a function. If $\alpha : [a, b] \to \mathbb{R}$ is a monotonically increasing function on $[a, b]$, then, for every $t_i \in [x_{i-1}, x_i]$, put

$$S(f, P, \alpha) = \sum_{i=1}^{n} f(t_i)(\alpha(x_i) - \alpha(x_{i-1})),$$

which is called the *Riemann-Stieltjes sum* of f with respect to P and α.

Note that, if $\alpha(x) = x$ for every $x \in [a, b]$, then we have the Riemann sum, that is,

$$S(f, P) = \sum_{i=1}^{n} f(t_i)(x_i - x_{i-1}).$$

Example 8.6.1. Let $P \in \wp[a, b]$ and $\alpha(x) = 3x$. Then, we have

$$S(f, P, \alpha) = 3S(f, P).$$

Definition 8.6.2. Let $f : [a, b] \to \mathbb{R}$ be a bounded function and $\alpha : [a, b] \to \mathbb{R}$ be a monotonically increasing function on $[a, b]$. Let $P = \{a = x_0, x_1, x_2, \ldots, x_n = b\}$ be a partition of $[a, b]$. For each subinterval $I_i = [x_{i-1}, x_i]$, put

$$M_i = \sup\{f(x) : x \in [x_{i-1}, x_i]\}, \quad m_i = \inf\{f(x) : x \in [x_{i-1}, x_i]\}.$$

Then, the *upper Riemann-Stieltjes sum* $U(f, P, \alpha)$ and the *lower Riemann-Stieltjes sum* $L(f, P, \alpha)$ are defined as follows, respectively:

$$U(f, P, \alpha) = \sum_{i=1}^{n} M_i(\alpha(x_i) - \alpha(x_{i-1})) = \sum_{i=1}^{n} M_i \Delta\alpha_i$$

and

$$L(f, P, \alpha) = \sum_{i=1}^{n} m_i(\alpha(x_i) - \alpha(x_{i-1})) = \sum_{i=1}^{n} m_i \Delta\alpha_i.$$

From the definitions of $L(f, P, \alpha)$ and $U(f, P, \alpha)$, we have the following.

Theorem 8.6.1. *Let $f : [a, b] \to \mathbb{R}$ be a bounded function and $\alpha : [a, b] \to \mathbb{R}$ be a monotonically increasing function on $[a, b]$. If $P, Q \in \wp[a, b]$ with $P \subset Q$. Then, we have*

$$L(f, P, \alpha) \leq L(f, Q, \alpha), \quad U(f, P, \alpha) \geq U(f, Q, \alpha).$$

Theorem 8.6.2. *Let $f : [a, b] \to \mathbb{R}$ be a bounded function and $\alpha : [a, b] \to \mathbb{R}$ be a monotonically increasing function on $[a, b]$. Then, for every $P, Q \in \wp[a, b]$, we have*

$$L(f, P, \alpha) \leq Q(f, Q, \alpha).$$

Proof. For any $P, Q \in \wp[a, b]$, we have $P \subset P \cup Q$ and $Q \subset P \cup Q$ and so

$$L(f, P, \alpha) \leq L(f, P \cup Q, \alpha) \leq U(f, P \cup Q, \alpha) \leq U(f, Q, \alpha).$$

Definition 8.6.3. Let $f : [a, b] \to \mathbb{R}$ be a bounded function and $\alpha : [a, b] \to \mathbb{R}$ be a monotonically increasing function on $[a, b]$.

(1) The *upper Riemann-Stieltjes integral* of f on $[a, b]$ is the number

$$\overline{\int_a^b} f(x) d\alpha(x) = \inf\{U(f, P, \alpha) : P \in \wp[a, b]\};$$

(2) The *lower Riemann-Stieltjes integral* of f on $[a, b]$ is the number

$$\underline{\int_a^b} f(x) d\alpha(x) = \sup\{L(f, P, \alpha) : P \in \wp[a, b]\};$$

(3) f is said to be *Riemann-Stieltjes integrable* on $[a, b]$ if

$$\overline{\int_a^b} f(x) d\alpha(x) = \underline{\int_a^b} f(x) d\alpha(x);$$

(4) In this case, the *Riemann-Stieltjes integral* of f is defined to be the number

$$\overline{\int_a^b} f(x) d\alpha(x) = \underline{\int_a^b} f(x) d\alpha(x),$$

which is denoted by

$$\int_a^b f(x) d\alpha(x);$$

(5) f is said to be *not Riemann integrable* on $[a, b]$ if

$$\overline{\int_a^b} f(x) d\alpha(x) \neq \underline{\int_a^b} f(x) d\alpha(x).$$

Note that $\Re_\alpha[a, b]$ denotes the set of Riemann-Stieltjes integrable functions on $[a, b]$.

Example 8.6.2. Let $\alpha(x) = c$ for all $x \in [a, b]$, where c is a constant, and $f : [a, b] \to \mathbb{R}$ be a bounded function on $[a, b]$. Then, we have

$$\int_a^b f(x) d\alpha(x) = 0.$$

Solution. Let $P = \{a = x_0, x_1, x_2, \ldots, x_n = b\}$ be a partition of $[a, b]$. Since $\alpha(x) = c$ for every $x \in [a, b]$, we have $U(f, P, \alpha) = L(f, P, \alpha)$ and so

$$\overline{\int_a^b} f(x)d\alpha(x) = 0 = \underline{\int_a^b} f(x)d\alpha(x).$$

Thus, $f \in \Re_\alpha[a, b]$ and

$$\int_a^b f(x)d\alpha(x) = 0.$$

This completes the proof.

Theorem 8.6.3. (The Riemann-Stieltjes Integral Test) *Let $f : [a, b] \to \mathbb{R}$ be a bounded function and $\alpha : [a, b] \to \mathbb{R}$ be a monotonically increasing function on $[a, b]$. Then, the following are equivalent:*

(1) $f \in \Re_\alpha[a, b]$;

(2) *For every $\varepsilon > 0$, there exists $P \in \wp[a, b]$ such that*

$$U(f, P, \alpha) - L(f, P, \alpha) < \varepsilon.$$

Proof. (1) $\Longrightarrow$ (2) Suppose that $f \in \Re_\alpha[a, b]$. Since

$$\int_a^b f(x)d\alpha(x) = \inf\{U(f, P, \alpha) : P \in \wp[a, b]\},$$

for every $\varepsilon > 0$, $\int_a^b f(x)dx + \frac{\varepsilon}{2}$ is not a lower bound of $\{U(f, P, \alpha) : P \in \wp[a, b]\}$, and so there exists $P_1 \in \wp[a, b]$ such that

$$U(f, P_1, \alpha) < \int_a^b f(x)d\alpha(x) + \frac{\varepsilon}{2}.$$

Also, since

$$\int_a^b f(x)d\alpha(x) = \inf\{L(f, P, \alpha) : P \in \wp[a, b]\},$$

$\int_a^b f(x)d\alpha(x) - \frac{\varepsilon}{2}$ is not an upper bound of $\{L(f, P, \alpha) : P \in \wp[a, b]\}$, and so there exists $P_2 \in \wp[a, b]$ such that

$$\int_a^b f(x)d\alpha(x) - \frac{\varepsilon}{2} < L(f, P_2, \alpha).$$

Let $P = P_1 \cup P_2$. Then, we have

$$L(f, P_2, \alpha) \le L(f, P, \alpha) \le U(f, P, \alpha) \le U(f, P_1, \alpha).$$

Thus, we have

$$U(f, P, \alpha) - L(f, P, \alpha) \le U(f, P_1, \alpha) - L(f, P_2, \alpha)$$

$$< \left(\int_a^b f(x)d\alpha(x) + \frac{\varepsilon}{2}\right) - \left(\int_a^b f(x)d\alpha(x) - \frac{\varepsilon}{2}\right) = \varepsilon.$$

(2) $\implies$ (1) Suppose that, for every $\varepsilon > 0$, there exists $P \in \wp[a, b]$ such that

$$U(f, P, \alpha) - L(f, P, \alpha) < \varepsilon.$$

Since $\overline{\int_a^b} f(x) d\alpha(x) \le U(f, P, \alpha)$ and $L(f, P, \alpha) \le \underline{\int_a^b} f(x) d\alpha(x)$, we have

$$0 \le \overline{\int_a^b} f(x) d\alpha(x) - \underline{\int_a^b} f(x) d\alpha(x) \le U(f, P, \alpha) - L(f, P, \alpha) < \varepsilon$$

and so

$$\left| \overline{\int_a^b} f(x) d\alpha(x) - \underline{\int_a^b} f(x) d\alpha(x) \right| = \overline{\int_a^b} f(x) d\alpha(x) - \underline{\int_a^b} f(x) d\alpha(x) < \varepsilon.$$

Therefore, we have

$$\overline{\int_a^b} f(x) d\alpha(x) = \underline{\int_a^b} f(x) d\alpha(x)$$

and $f \in \Re_\alpha[a, b]$. This completes the proof.

Example 8.6.3. Let $f, \alpha : [0, 2] \to \mathbb{R}$ by

$$f(x) = \begin{cases} 0, & \text{if } x \in [0, 1], \\ 1, & \text{if } x \in (1, 2], \end{cases}$$

and

$$\alpha(x) = \begin{cases} 0, & \text{if } x \in [0, 1), \\ 1, & \text{if } x \in [1, 2], \end{cases}$$

respectively. Show that f is Riemann-Stieltjes integrable on $[0, 2]$ with respect to α.

Solution. Let $P = \{0 = x_0, x_1, x_2, \ldots, x_{j-1}, 1, x_{j+1}, \ldots, x_n = 2\}$ be a partition of $[0, 2]$, where $x_j = 1$. Then, we have

$$\begin{aligned} U(f, P, \alpha) &= \sum_{i=1}^n M_i(\alpha(x_i) - \alpha(x_{i-i})) \\ &= \sum_{i=1}^{j-1} M_i(\alpha(x_i) - \alpha(x_{i-1})) + M_j(\alpha(x_j) - \alpha(x_{j-1})) \\ &\quad + \sum_{i=j+1}^n M_i(\alpha(x_i) - \alpha(x_{i-1})) \\ &= M_j(\alpha(x_j) - \alpha(x_{j-1})) \\ &= M_j = \sup_{x \in [x_j, 1]} f(x) = 0. \end{aligned}$$

Similarly, we can have

$$L(f, P, \alpha) = m_j(\alpha(x_j) - \alpha(x_{j-1})) = m_j = \inf_{x \in [x_j, 1]} f(x) = 0.$$

Therefore, we ahve

$$U(f, P, \alpha) = 0 = L(f, P, \alpha).$$

Therefore, we have $f \in \Re_\alpha[0, 2]$ and $\int_0^2 f(x) d\alpha(x) = 0$.

Theorem 8.6.4. *Let $f : [a, b] \to \mathbb{R}$ be a continuous function and $\alpha : [a, b] \to \mathbb{R}$ be a monotonically increasing function on $[a, b]$. Then, $f \in \Re_\alpha[a, b]$.*

Proof. If $\alpha(x) = \alpha(y)$ for every $x, y \in [a, b]$, that is, $\alpha(x)$ is a constant function, then, for every $P \in \wp[a, b]$, we have

$$U(f, P, \alpha) - L(f, P, \alpha) = \sum_{i=1}^{n} (M_i - m_i) \Delta \alpha_i = 0$$

and so $f \in \Re_\alpha[a, b]$.

Let $\alpha(a) < \alpha(b)$. Since $f : [a, b] \to$ is uniformly continuous on $[a, b]$, for every $\varepsilon > 0$, there exists $\delta > 0$ such that

$$|s - t| < \delta, \ s, t \in [a, b] \implies |f(s) - f(t)| < \frac{\varepsilon}{\alpha(b) - \alpha(a)}.$$

Let $P = \{a = x_0, x_1, x_2, \ldots, x_n = b\}$ be a partition of $[a, b]$ with $\Delta x_i = |x_i - x_{i-1}| < \delta$. Then, we have

$$M_i - m_i \le \frac{\varepsilon}{\alpha(b) - \alpha(a)}$$

and so

$$U(f, P, \alpha) - L(f, P, \alpha) = \sum_{i=1}^{n} (M_i - m_i) \Delta \alpha_i$$

$$\le \frac{\varepsilon}{\alpha(b) - \alpha(a)} \sum_{i=1}^{n} \Delta \alpha_i$$

$$= \varepsilon.$$

Therefore, we have $f \in \Re_\alpha[a, b]$. This completes the proof.

Theorem 8.6.5. *Let $f : [a, b] \to \mathbb{R}$ be a monotone function and $\alpha : [a, b] \to \mathbb{R}$ be a continuous and monotonically increasing function on $[a, b]$. Then, $f \in \Re_\alpha[a, b]$.*

Proof. For every $n \ge 1$, let $P \in \wp[a, b]$ with

$$\Delta \alpha_i = \alpha(x_j) - \alpha(x_{j-1}) = \frac{\alpha(b) - \alpha(a)}{n}.$$

Since α is continuous on $[a, b]$, by the Intermediate Value Theorem, there exists $x_1 \in [x_0, x_1]$ such that

$$\alpha(x_1) = \alpha(a) + \frac{\alpha(b) - \alpha(a)}{n}.$$

Inductively, there exists $x_k \in [x_{k-1}, b]$ such that

$$\alpha(x_k) = \alpha(a) + k\frac{\alpha(b) - \alpha(a)}{n}.$$

Thus, we can consider a partition $P = \{a = x_0, x_1, x_2, \ldots, x_n = b\}$ with

$$\Delta\alpha_i = \alpha(x_j) - \alpha(x_{j-1}) = \frac{\alpha(b) - \alpha(a)}{n}.$$

Let $f : [a, b] \to \mathbb{R}$ be a monotonically increasing function. Then, we have

$$M_i = f(x_i), \quad m_i = f(x_{i-1}) \text{ for every } i = 1, 2, \ldots, n.$$

Thus, we have

$$\begin{aligned}
U(f, P, \alpha) - L(f, P, \alpha) &= \sum_{i=1}^{n}(M_i - m_i)\Delta\alpha_i \\
&= \frac{\alpha(b) - \alpha(a)}{n}\sum_{i=1}^{n}(f(x_i) - f(x_{i-1})) \\
&= \frac{\alpha(b) - \alpha(a)}{n}(f(b) - f(a)).
\end{aligned}$$

Therefore, for every $\varepsilon > 0$, taking a sufficiently large n such that

$$\frac{(\alpha(b) - \alpha(a))(f(b) - f(a))}{\varepsilon} < n,$$

we have

$$U(f, P, \alpha) - L(f, P, \alpha) < \varepsilon$$

and so $f \in \mathfrak{R}_\alpha[a, b]$. This completes the proof.

Theorem 8.6.6. *Let $f : [a, b] \to \mathbb{R}$ be a bounded function and $\alpha : [a, b] \to \mathbb{R}$ be a monotonically increasing function on $[a, b]$. Suppose that there exist constants M and m such that $m \le f(x) \le M$ for all $x \in [a, b]$. If $f \in \mathfrak{R}_\alpha[a, b]$ and $g : [m, M] \to \mathbb{R}$ is monotonically increasing on $[m, M]$, then $g \circ f \in \mathfrak{R}_\alpha[a, b]$.*

Proof. We can prove this theorem as in the proof of Theorem 8.2.1. So, we omit the proof.

The following theorem shows some relations between the Riemann integral and the Riemann-Stieltjes integral:

Theorem 8.6.7. *Let $f : [a, b] \to \mathbb{R}$ be Riemann integrable on $[a, b]$ and $\alpha : [a, b] \to \mathbb{R}$ be a monotonically increasing function on $[a, b]$. If $\alpha' \in \Re[a, b]$, then we have the following:*

(1) $f \in \Re_\alpha[a, b]$;

(2) *If $f\alpha' \in \Re[a, b]$, then*

$$\int_a^b f(x)d\alpha(x) = \int_a^b f(x)\alpha'(x)dx.$$

Proof. (1) Since $f, \alpha' \in \Re[a, b]$, we have $f\alpha' \in \Re[a, b]$. Let $P = \{a = x_0, x_1, x_2, \ldots, x_n = b\}$ be a partition of $[a, b]$, and, for each subinterval $[x_{i-1}, x_i]$ $(i = 1, 2, \ldots, n)$, if we use the Mean Value Theorem for the function α, then there exists $t_i \in (x_{i-1}, x_i)$ such that

$$\alpha(x_i) - \alpha(x_{i-1}) = \alpha'(t_i)(x_i - x_{i-1}).$$

Since $\alpha' \in \Re[a, b]$, α' is bounded, and so, for every $x \in [a, b]$, there exists $K > 0$ such that $|\alpha'(x)| \leq K$. Thus, we have

$$U(f, P, \alpha) - L(f, P, \alpha) = \sum_{i=1}^n (M_i - m_i)(\alpha(x_i) - \alpha(x_{i-1}))$$

$$= \sum_{i=1}^n (M_i - m_i)\alpha'(t_i)(x_i - x_{i-1})$$

$$\leq K \sum_{i=1}^n (M_i - m_i)(x_i - x_{i-1})$$

$$= K(U(f, P) - L(f, P)).$$

On the other hand, since $f \in \Re[a, b]$, for every $\varepsilon > 0$, there exists $P(\varepsilon) \in \wp[a, b]$ such that

$$U(f, P(\varepsilon)) - L(f, P(\varepsilon)) < \frac{\varepsilon}{K}.$$

Therefore, we have

$$U(f, P(\varepsilon), \alpha) - L(f, P(\varepsilon), \alpha) \leq K(U(f, P(\varepsilon)) - L(f, P(\varepsilon))) < \varepsilon,$$

that is, $f \in \Re_\alpha[a, b]$.

(2) Let $\varepsilon > 0$. Since $f\alpha' \in \Re[a, b]$, there exists a partiton $P = \{a = x_0, x_1, x_2, \ldots, x_n = b\}$ of $[a, b]$ such that, for any $s_i \in [x_{i-1}, x_i]$,

$$\int_a^b f(x)\alpha'(x)dx - \varepsilon < \sum_{i=1}^n (f\alpha')(s_i)\Delta x_i < \int_a^b f(x)\alpha'(x)dx + \varepsilon.$$

Further, since $\alpha'(x) \geq 0$ for every $x \in [a, b]$, we have

$$U(f, P, \alpha) = \sum_{i=1}^{n} M_i(\alpha(x_i) - \alpha(x_{i-1}))$$

$$= \sum_{i=1}^{n} M_i \alpha'(t_i)(x_i - x_{i-1})$$

$$\geq \sum_{i=1}^{n} f(t_i) \Delta x_i$$

$$> \int_a^b f(x) \alpha'(x) dx - \varepsilon.$$

Since $\varepsilon > 0$ is arbitrary, we have

$$\int_a^b f(x) d\alpha(x) \geq \int_a^b f(x) \alpha'(x) dx.$$

Similarly, we have

$$L(f, P, \alpha) = \sum_{i=1}^{n} m_i(\alpha(x_i) - \alpha(x_{i-1}))$$

$$= \sum_{i=1}^{n} m_i \alpha'(t_i)(x_i - x_{i-1})$$

$$\leq \sum_{i=1}^{n} f(t_i) \Delta x_i$$

$$< \int_a^b f(x) \alpha'(x) dx + \varepsilon.$$

Since $\varepsilon > 0$ is arbitrary, we have

$$\int_a^b f(x) d\alpha(x) \leq \int_a^b f(x) \alpha'(x) dx.$$

Therefore, we have

$$\int_a^b f(x) d\alpha(x) = \int_a^b f(x) \alpha'(x) dx.$$

This completes the proof.

Example 8.6.4. Evaluate $\int_0^1 x^2 dx^2$.

Solution. Put $\alpha(x) = x^2$, then we have $\alpha'(x) = 2x$ and so

$$\int_0^1 x^2 dx^2 = \int_0^1 x^2 2x dx = 2 \int_0^1 x^3 dx = \frac{1}{2}.$$

8.7 Functions of Bounded Variation

In this section, we give some properties of function of bounded varition and show that, if $f : [a, b] \to \mathbb{R}$ is continuous on $[a, b]$ and $\alpha : [a, b] \to \mathbb{R}$ is a function of bounded variation on $[a, b]$, then f is Riemann-Stieltjes integrable on $[a, b]$. Also, we introduce the Integration by Parts for the Riemann-Stieltjes integrals.

Definition 8.7.1. (1) Let $\alpha : [a, b] \to \mathbb{R}$ be a function and $P = \{a = x_0, x_1, x_2, \ldots, x_n = b\}$ be a partition of $[a, b]$. Then,

$$V(\alpha, P) = \sum_{i=1}^{n} |\alpha(x_i) - \alpha(x_{i-1})|$$

is called the *variation* of α for P;

(2) If the set

$$K = \{V(\alpha, P) : P \in \wp[a, b]\}$$

is bounded from above, then α is called a *function of bounded variation*, which is denoted by $\alpha \in BV[a, b]$;

(3) If the set

$$K = \{V(\alpha, P) : P \in \wp[a, b]\}$$

is bounded from above, then

$$V_a^b(\alpha) = \sup\{V(\alpha, P) : P \in \wp[a, b]\}$$

is called the *total variation* of α on $[a, b]$.

Theorem 8.7.1. *Let $\alpha : [a, b] \to \mathbb{R}$ be a monotone function and $\alpha \in BV[a, b]$. Then, we have*

$$V_a^b(\alpha) = |\alpha(b) - \alpha(a)|.$$

Proof. Let $\alpha : [a, b] \to \mathbb{R}$ be increasing on $[a, b]$ and

$$P = \{a = x_0, x_1, x_2, \ldots, x_n = b\}$$

be a partition of $[a, b]$. Then, we have

$$V(\alpha, P) = \sum_{i=1}^{n} (\alpha(x_i) - \alpha(x_{i-1})) = \alpha(b) - \alpha(a).$$

Thus, for every $P \in \wp[a, b]$, since $K = \alpha(b) - \alpha(a)$, we have

$$V_a^b(\alpha) = \alpha(b) - \alpha(a).$$

Similarly, if $\alpha : [a, b] \to \mathbb{R}$ is decreasing on $[a, b]$, then we can get the same result. This completes the proof.

Theorem 8.7.2. *If $\alpha : [a, b] \to \mathbb{R}$ is a function of bounded variation on $[a, b]$, that is, $\alpha \in BV[a, b]$, then α is bounded on $[a, b]$.*

Proof. For every $x \in [a, b]$, let $P = \{a, x, b\}$ be a partition of $[a, b]$. Then, we have

$$V(\alpha, P) = |\alpha(b) - \alpha(x) + |\alpha(x) - \alpha(a)|$$

and

$$|\alpha(x)| - |\alpha(b)| + |\alpha(x)| - |\alpha(a)| \leq |\alpha(x) - \alpha(b) + |\alpha(x) - \alpha(a)|$$
$$= V(\alpha, P) \leq V_a^b(\alpha).$$

Thus, we have

$$|\alpha(x)| \leq \frac{1}{2}(V_a^b(\alpha) + |\alpha(a)| + |\alpha(b)|).$$

Put $M = \max\{\alpha(a), \alpha(b)\}$, it follows that, for every $x \in [a, b]$,

$$|\alpha(x)| \leq \frac{1}{2}V_a^b(\alpha) + M$$

and so f is bounded on $[a, b]$. This completes the proof.

Example 8.7.1. Let $\alpha : [0, 1] \to \mathbb{R}$ be a function defined by

$$\alpha(x) = \begin{cases} x \sin \frac{1}{x}, & \text{if } x \neq 0, \\ 0, & \text{if } x = 0. \end{cases}$$

Show $\alpha \notin BV[0, 1]$.

Solution. Let $P = \{0 = x_0, x_1, x_2, \ldots, x_i, \ldots, x_n = 1\}$ be a partition of $[0, 1]$, where

$$x_i = \frac{2}{(2(n - i) + 1)\pi} \quad \text{for every } i = 0, 1, 2, \ldots, n.$$

Then, we have

$$V(\alpha, P) = \sum_{i=1}^{n} |\alpha(x_i) - \alpha(x_{i-1})|$$

$$= |\alpha(x_1) - \alpha(x_0)| + |\alpha(x_n) - \alpha(x_{n-1})| + \sum_{i=2}^{n-1} |\alpha(x_i) - \alpha(x_{i-1})|$$

$$= \left| x_1 \sin \frac{1}{x_1} - 0 \right| + \left| \sin 1 - x_{n-1} \sin \frac{1}{x_{n-1}} \right|$$

$$+ \sum_{i=2}^{n-1} \left| x_i \sin \frac{1}{x_i} - x_{i-1} \sin \frac{1}{x_{i-1}} \right|$$

$$= \frac{2}{(2n-1)\pi} + \left| \sin 1 - \frac{2}{3\pi} \right|$$

$$+ \sum_{i=2}^{n-1} \left[\frac{2}{(2(n-i)+1)\pi} + \frac{2}{(2(n-i)+3)\pi} \right]$$

$$\geq \frac{2}{\pi} \sum_{i=2}^{n-1} \left[\frac{1}{2(n-i)+1} + \frac{1}{2(n-i)+3} \right]$$

$$= \frac{2}{\pi} \sum_{k=1}^{n-2} \left[\frac{1}{2k+1} + \frac{1}{2k+3} \right]$$

and so

$$V(\alpha, P) \geq \frac{4}{\pi} \sum_{k=1}^{n-2} \frac{1}{2k+3}.$$

Since $\frac{4}{\pi} \sum_{k=1}^{n-2} \frac{1}{2k+3} \to \infty$ as $n \to \infty$, we have $V(\alpha, P) \to \infty$, and so, by Theorem 8.7.2, α is not a function of bounded variation, that is, $\alpha \notin BV[0,1]$.

Theorem 8.7.3. *Let* $\alpha : [a, b] \to \mathbb{R}$ *be a continuous function on* $[a, b]$ *and* $\alpha : (a, b) \to \mathbb{R}$ *be differentiable on* (a, b)*. If* α' *is bounded on* (a, b)*, then* $\alpha \in BV[a, b]$*.*

Proof. Since α' is bounded on (a, b), there exists $M > 0$ such that

$$|\alpha'(x)| < M \quad \text{for every} \quad x \in (a, b).$$

Let $P = \{a = x_0, x_1, x_2, \ldots, x_n = b\}$ be a partition of $[a, b]$. Then, by Lagrange's Mean Value Theorem, for each $i = 1, 2, \ldots, n$, there exists $t_i \in (x_{i-1}, x_i)$ such that

$$\Delta \alpha_i = \alpha(x_i) - \alpha(x_{i-1}) = \alpha'(t_i)(x_i - x_{i-1}).$$

Thus, we have

$$V(\alpha, P) = \sum_{i=1}^{n} (\alpha(x_i) - \alpha(x_{i-1}))$$

$$= \sum_{i=1}^{n} |\alpha'(t_i)|(x_i - x_{i-1})$$

$$< M \sum_{i=1}^{n} (x_i - x_{i-1})$$

$$= M(b - a)$$

and so $\alpha \in BV[a, b]$. This completes the proof.

Example 8.7.2. Let $\alpha : [-1, 1] \to \mathbb{R}$ be a function defined by

$$\alpha(x) = \begin{cases} x^2 \sin \frac{1}{x}, & \text{if } x \neq 0, \\ 0, & \text{if } x = 0. \end{cases}$$

Show $\alpha \in BV[-1, 1]$.

Solution. α is continuous on $[-1, 1]$, α is differentiable on $(-1, 1)$ and

$$
\alpha'(x) = \begin{cases} x^2 \sin \frac{1}{x} - \cos \frac{1}{x}, & \text{if } x \neq 0, \\ 0, & \text{if } x = 0. \end{cases}
$$

Thus, we have $|\alpha'(x)| \leq 3$. Therefore, by Theorem 8.7.3, $\alpha \in BV[-1, 1]$.

Theorem 8.7.4. *Let $\alpha, \beta \in BV[a, b]$ and $c \in \mathbb{R}$. Then, we have the following:*

(1) $\alpha + \beta \in BV[a, b]$ and $V_a^b(\alpha + \beta) \leq V_a^b(\alpha) + V_a^b(\beta)$;

(2) $c\alpha \in BV[a, b]$ and $V_a^b(c\alpha) = |c| V_a^b(\alpha)$;

(3) $\alpha - \beta \in BV[a, b]$.

Proof. (1) Let $P = \{a = x_0, x_1, x_2, \ldots, x_n = b\}$ be a partition of $[a, b]$. Then, we have

$$
\sum_{i=1}^n |\alpha(x_i) + \beta(x_i) - (\alpha(x_{i-1}) + \beta(x_{i-1}))|
$$

$$
\leq \sum_{i=1}^n |\alpha(x_i) + \alpha(x_{i-1})| + \sum_{i=1}^n |\beta(x_i) + \beta(x_{i-1})|
$$

$$
\leq V_a^b(\alpha) + V_a^b(\beta)
$$

and so $\alpha + \beta \in BV[a, b]$ and

$$
V_a^b(\alpha + \beta) \leq V_a^b(\alpha) + V_a^b(\beta).
$$

(2) From

$$
\sum_{i=1}^n |(c\alpha)(x_i) + (c\alpha)(x_{i-1})| = |c| \sum_{i=1}^n |\alpha(x_i) + \alpha(x_{i-1})|,
$$

we have $c\alpha \in BV[a, b]$ and

$$
V_a^b(c\alpha) = |c| V_a^b(\alpha).
$$

(3) Since $\alpha - \beta = \alpha + (-1)\beta$, by (1) and (2), we have $\alpha - \beta \in BV[a, b]$. This completes the proof.

Theorem 8.7.5. *The following are equivalent:*

(1) $\alpha \in BV[a, b]$;

(2) $\alpha = \beta - \gamma$, where $\beta, \gamma : [a, b] \to \mathbb{R}$ are monotonically increasing on $[a, b]$.

Proof. $(1) \implies (2)$ Let $\alpha \in BV[a, b]$. Define a function $\beta : [a, b] \to \mathbb{R}$ by

$$\beta(x) = \begin{cases} V_a^x(\alpha), & \text{if } x \in (a, b], \\ 0, & \text{if } x = a, \end{cases}$$

and put $\gamma = \beta - \alpha$. For every $x, y \in [a, b]$ with $x < y$, since

$$\beta(y) - \beta(x) = V_a^y(\alpha) - V_a^x(\alpha) = V_x^y(\alpha) \geq 0,$$

β is monotonically increasing on $[a, b]$. On the other hand, since $\{x, y\}$ is a partition of $[x, y]$, we have

$$\alpha(y) - \alpha(x) \leq V_x^y(\alpha)$$

and

$$\gamma(y) - \gamma(x) = V_x^y(\alpha) - (\alpha(y) - \alpha(x)).$$

Therefore, γ is monotonically increasing on $[a, b]$ and $\alpha = \beta - \gamma$.

$(2) \implies (1)$ Let $\beta, \gamma : [a, b] \to \mathbb{R}$ be monotonically increasing on $[a, b]$. Then, by Theorem 8.7.1, $\beta, \gamma \in \in BV[a, b]$. Thus, by Theorem 8.7.4, we have $\alpha = \beta - \gamma \in BV[a, b]$. This completes the proof.

As an application of Theorem 8.7.5, we have the following:

Theorem 8.7.6. *If $\alpha \in BV[a, b]$, then the following are equivalent:*

(1) *α is continuous on (a, b);*

(2) *α is uniformly continuous on (a, b).*

Proof. $(2) \implies (1)$ If α is uniformly continuous on (a, b), then clearly α is continuous on (a, b).

$(1) \implies (2)$ Suppose that α is continuous on (a, b). By Theorem 8.7.5, $\alpha = \beta - \gamma$, where $\beta, \gamma : [a, b] \to \mathbb{R}$ are monotonically increasing on (a, b). Putting

$$M_1 = \sup\{\beta(x) : x < b\}, \quad M_2 = \sup\{\gamma(x) : x < b\}$$

and

$$m_1 = \inf\{\beta(x) : a < x\}, \quad m_2 = \inf\{\gamma(x) : a < x\},$$

we have

$$\lim_{x \to a^+} \beta(x) = m_1, \quad \lim_{x \to a^+} \gamma(x) = m_2$$

and

$$\lim_{x \to b^-} \beta(x) = M_1, \quad \lim_{x \to b^-} \gamma(x) = M_2.$$

Now, if we define a function $f : [a, b] \to \mathbb{R}$ by

$$f(x) = \begin{cases} \alpha(x), & \text{if } x \in (a, b), \\ m_1 - m_2, & \text{if } x = a, \\ M_1 - M_2, & \text{if } x = b, \end{cases}$$

then $f : [a, b] \to \mathbb{R}$ is continuous on $[a, b]$, and so $f : [a, b] \to \mathbb{R}$ is uniformly continuous on $[a, b]$. Thus, f is uniformly continuous on (a, b), and so, since $\alpha = f$ on (a, b), α is uniformly continuous on (a, b). This completes the proof.

Theorem 8.7.7. *Let $\alpha \in BV[a, b]$ and $c \in (a, b)$. Then, we have the following:*

(1) $\alpha \in BV[a, c] \cap BV[c, b]$;

(2) $V_a^b(\alpha) = V_a^c(\alpha) + V_c^b(\alpha)$.

Proof. Let $P = \{a = x_0, x_1, x_2, \ldots, x_n = c\}$ be a partition of $[a, c]$ and $Q = \{c = y_0, y_1, y_2, \ldots, y_m = b\}$ be a partition of $[c, b]$. Then, $P \cup Q$ is a partition of $[a, b]$ and

$$\sum_{i=1}^{n} |\alpha(x_i) + \alpha(x_{i-1})| + \sum_{i=1}^{m} |\alpha(y_i) + \alpha(y_{i-1})| \leq V_a^b(\alpha).$$

Since

$$V(\alpha, P) = \sum_{i=1}^{n} |\alpha(x_i) + \alpha(x_{i-1})| \leq V_a^b(\alpha) - \sum_{i=1}^{m} |\alpha(y_i) + \alpha(y_{i-1})|,$$

we have $\alpha \in BV[a, c]$ and

$$V_a^c(\alpha) \leq V_a^b(\alpha) - \sum_{i=1}^{m} |\alpha(y_i) + \alpha(y_{i-1})|.$$

Similarly, for the partition Q of $[c, b]$, we have

$$V(\alpha, Q) = \sum_{i=1}^{m} |\alpha(y_i) + \alpha(y_{i-1})| \leq V_a^b(\alpha) - V_a^c(\alpha)$$

and so $\alpha \in BV[c, b]$ and $V_c^b(\alpha) \leq V_a^b(\alpha) - V_a^c(\alpha)$, that is,

$$V_a^c(\alpha) + V_c^b(\alpha) \leq V_a^b(\alpha). \tag{8.7.1}$$

On the other hand, let $P = \{a = x_0, x_1, x_2, \ldots, x_n = c\}$ be a partition of $[a, b]$. Then, $P \cup \{c\} = \{y_0, y_1, y_2, \ldots, y_m\}$ be a partition of $[a, b]$, and, for some j with $0 \leq j \leq m$, we have $y_j = c$. Then, $\{y_0, y_1, y_2, \ldots, y_j\}$ is a partition of $[a, c]$ and $\{y_j, y_{j+1}, y_{j+2}, \ldots, y_m\}$ is a partition of $[c, b]$, and so we have

$$\sum_{i=1}^{n} |\alpha(x_i) + \alpha(x_{i-1})| \leq \sum_{i=1}^{j} |\alpha(y_i) + \alpha(y_{i-1})| + \sum_{i=j+1}^{m} |\alpha(x_i) + \alpha(x_{i-1})|$$

$$\leq V_a^c(\alpha) + V_c^b(\alpha).$$

Thus, we have

$$V_a^b(\alpha) \le V_a^c(\alpha) + V_c^b(\alpha). \tag{8.7.2}$$

Therefore, from (8.7.1) and (8.7.2), it follows that

$$V_a^b(\alpha) = V_a^c(\alpha) + V_c^b(\alpha).$$

This completes the proof.

Theorem 8.7.8. *Let $f : [a, b] \to \mathbb{R}$ be a bounded function and $\alpha : [a, b] \to \mathbb{R}$ be a monotonically increasing function on $[a, b]$. Then, the following are equivalent:*

(1) *$f \in \mathcal{R}_\alpha[a, b]$;*

(2) *For any $\varepsilon > 0$, there exists a partition $P \in \wp[a, b]$ such that, for some $t \in \mathbb{R}$, if Q is a partition of $[a, b]$ with $P \subset Q$,*

$$|S(f, Q, \alpha) - t| < \varepsilon,$$

where $t = \int_a^b f(x)d\alpha(x)$.

Proof. (1) $\implies$ (2) Let $f \in \mathcal{R}_\alpha[a, b]$. Then, for every $\varepsilon > 0$, there exists a partition $P \in \wp[a, b]$ such that

$$U(f, P, \alpha) - L(f, P, \alpha) < \varepsilon.$$

If $P \subset Q$, then we have

$$U(f, Q, \alpha) - L(f, Q, \alpha) \le U(f, P, \alpha) - L(f, P, \alpha) < \varepsilon.$$

On the other hand, since

$$L(f, Q, \alpha) \le \int_a^b f(x)d\alpha(x) \le U(f, Q, \alpha),$$

we have

$$\left| S(f, Q, \alpha) - \int_a^b f(x)d\alpha(x) \right| < \varepsilon,$$

where we put $t = \int_a^b f(x)d\alpha(x)$.

(2) $\implies$ (1) If $\alpha(b) = \alpha(a)$, then, since $\alpha(x)$ is a constant function, $f \in \mathcal{R}_\alpha[a, b]$ and $\int_a^b f(x)d\alpha(x) = 0$.

Let $\alpha(b) \ne \alpha(a)$ and $\varepsilon > 0$. Let $P = \{a = x_0, x_1, x_2, \ldots, x_n = b\}$ be a partition of $[a, b]$ such that

$$|S(f, Q, \alpha) - t| < \frac{\varepsilon}{4},$$

and put

$$M_i = \sup\{f(x) : x \in [x_{i-1} - x_i]\}$$

for every $i = 1, 2, \ldots, n$. Then, there exists $t_i \in [x_{i-1} - x_i]$ such that

$$M_i - \frac{\varepsilon}{4(\alpha(b) - \alpha(a))} < f(t_i)$$

and so

$$U(f, P, \alpha) - \frac{\varepsilon}{4} = \sum_{i=1}^{n} \left(M_i - \frac{\varepsilon}{4(\alpha(b) - \alpha(a))} \right) \leq \sum_{i=1}^{n} f(t_i) \Delta \alpha_i.$$

Put $T = \{t_1, t_2, \ldots, t_n\}$, where $t_i \in [x_{i-1} - x_i]$ for every $i = 1, 2, \ldots, n$. Then, we have

$$U(f, P, \alpha) - \frac{\varepsilon}{4} \leq S(f, P \cup T, \alpha).$$

Similarly, put $Q = \{q_1, q_2, \ldots, q_n\}$, where $q_i \in [x_{i-1} - x_i]$ for every $i = 1, 2, \ldots, n$. Then, we have

$$S(f, P \cup Q, \alpha) \leq L(f, P, \alpha) + \frac{\varepsilon}{4}.$$

Therefore, we have

$$U(f, P, \alpha) - L(f, P, \alpha) \leq \left[S(f, P \cup T, \alpha) + \frac{\varepsilon}{4} \right] + \left[-S(f, P \cup Q, \alpha) + \frac{\varepsilon}{4} \right]$$

$$\leq |S(f, P \cup T, \alpha) - t| + |S(f, P \cup Q, \alpha) - t| + \frac{\varepsilon}{2}$$

$$< \frac{\varepsilon}{4} + \frac{\varepsilon}{4} + \frac{\varepsilon}{2} = \varepsilon.$$

Therefore, we have $f \in \Re_\alpha[a, b]$. This completes the proof.

Theorem 8.7.9. (The Integration by Parts) *Let $f, \alpha : [a, b] \to \mathbb{R}$ be bounded on $[a, b]$. If $f \in \Re_\alpha[a, b]$, then we have the following:*

(1) $\alpha \in \Re_f[a, b]$;

(2) $\int_a^b f(x) d\alpha(x) = f(b)\alpha(b) - f(a)\alpha(a) - \int_a^b \alpha(x) df(x)$.

Proof. Let $\varepsilon > 0$ be arbitrary. Then, by Theorem 8.7.6, there exists a partition $P = \{a = x_0, x_1, x_2, \ldots, x_n = b\}$ of $[a, b]$ such that if Q is a partition of $[a, b]$ with $P \subset Q$,

$$\left| S(f, Q, \alpha) - \int_a^b f(x) d\alpha(x) \right| < \varepsilon. \tag{8.7.3}$$

On the other hand, for every $t_i \in [x_{i-1} - x_i]$ $(i = 1, 2, \ldots, n)$, if we put

$$S(f, P, \alpha) = \sum_{i=1}^{n} \alpha(t_i)(f(x_i) - f(x_{i-1})),$$

then we have

$$S(f, P, \alpha) = \sum_{i=1}^{n} \alpha(t_i)f(x_i) - \sum_{i=1}^{n} \alpha(x_i)f(x_i) - \sum_{i=1}^{n} \alpha(t_i)f(x_{i-1})$$

$$+ \sum_{i=1}^{n} \alpha(x_{i-1})f(x_{i-1}) + \alpha(x_n)f(x_n) - \alpha(x_0)f(x_0)$$

$$= - \sum_{i=1}^{n} f(x_i)(\alpha(x_i) - \alpha(t_i)) - \sum_{i=1}^{n} f(t_{i-1})(\alpha(t_i) - \alpha(x_{i-1}))$$

$$+ f(b)\alpha(b) - f(a)\alpha(a).$$

Putting $T = \{t_i, t_2, \ldots, t_n\}$, where $T-i \in\in [x_{i-1} - x_i]$ for every $i = 1, 2, \ldots, n$, we have

$$P \cup T = \{a = x_0, t_1, x_1, t_2, x_2, \ldots, x_{n-1}, t_n, x_n = b\}$$

and so

$$S(f, P, \alpha) = -S(f, P \cup T, \alpha) + f(b)\alpha(b) - f(a)\alpha(a)$$

$$= \left| \int_a^b f(x)d\alpha(x) - S(f, Q \cup T, \alpha) \right| < \varepsilon.$$

Thus, by (8.7.3), we have

$$\left| \int_a^b f(x)d\alpha(x) - S(f, P \cup T, \alpha) \right|$$

$$= \left| \int_a^b f(x)d\alpha(x) - f(b)\alpha(b) + f(a)\alpha(a) - S(f, P, \alpha) \right|$$

$$< \varepsilon.$$

Therefore, we have $\alpha \in \Re_f[a, b]$ and

$$\int_a^b f(x)d\alpha(x) = f(b)\alpha(b) - f(a)\alpha(a) - \int_a^b \alpha(x)df(x).$$

This completes the proof.

Corollary 8.7.10. *Let $f, \alpha, \beta : [a, b] \to \mathbb{R}$ be bounded on $[a, b]$, and let $c \in \mathbb{R}$. If $f \in \Re_\alpha[a, b] \cap \Re_\beta[a, b]$, then we have the following:*

(1) $f \in \Re_{\alpha+\beta}[a, b]$;

(2) $f \in \Re_{c\alpha}[a, b]$ *and* $\int_a^b f(x)d(c\alpha)(x) = \int_a^b cf(x)d\alpha(x)$.

Proof. (1) By Theorem 8.7.9, since $\alpha, \beta \in \Re_f[a, b]$, we have $\alpha + \beta \in \Re_f[a, b]$ and $f \in \Re_{\alpha+\beta}[a, b]$. Further, we have

$$\int_a^b f(x)d(\alpha + \beta)(x)$$

$$= f(b)(\alpha(b) + \beta(b)) - f(a)(\alpha(a) + \beta(a)) - \int_a^b (\alpha + \beta)(x)df(x)$$

$$= \left(f(b)\alpha(b) - f(a)\alpha(a) - \int_a^b \alpha(x)df(x) \right)$$

$$+ \left(f(b)\beta(b) - f(a)\beta(a) - \int_a^b \beta(x)df(x) \right)$$

$$= \int_a^b f(x)d\alpha(x) + \int_a^b f(x)d\beta(x).$$

(2) Similarly, we have

$$\int_a^b f(x)d(c\alpha)(x) = \int_a^b cf(x)d\alpha(x).$$

This completes the proof.

Theorem 8.7.11. *Let* $f : [a, b] \to \mathbb{R}$ *be a continuous function and* $\alpha \in BV[a, b]$. *Then,* $f \in \Re_\alpha[a, b]$.

Proof. Let $\beta, \gamma : [a, b] \to \mathbb{R}$ be monotonically increasing on $[a, b]$ with $\alpha = \beta - \gamma$. Then, we have

$$f \in \Re_\beta[a, b] \bigcap \Re_\gamma[a, b].$$

Thus, by Corollary 8.7.10, we have

$$f \in \Re_{\beta-\gamma}[a, b] = \Re_\alpha[a, b].$$

This completes the proof.

Theorem 8.7.12. *Let* $\alpha : [a, b] \to \mathbb{R}$ *be a continuous function and* $f \in BV[a, b]$. *Then,* $f \in \Re_\alpha[a, b]$.

8.8   Exercises

Excercise 8.1 Show that the following function

$$f(x) = \begin{cases} 0, & \text{if } 0 < x \leq 1, \\ 1, & \text{if } x = 0, \end{cases}$$

is Riemann integrable on $[0, 1]$ and

$$\int_0^1 f(x)dx = 0.$$

Excercise 8.2 Show that the following function

$$f(x) = \begin{cases} 0, & \text{if } x \in \mathbb{Q}, \\ 1, & \text{if } x \notin \mathbb{Q}, \end{cases}$$

is not Riemann integrable on any interval.

Hint. See Example 8.1.2.

Excercise 8.3 Show that $f(x) = x^3$ is Riemann integrable on any interval $[0, t]$, and, also, find

$$\int_0^t f(x)dx.$$

Ans. $\int_0^t f(x)dx = \frac{t^4}{4}$.

Excercise 8.4 Construct an example of a function f which is not Riemann integrable but whose $|f|$ is Riemann integrable.

Hint. Note that

$$f(x) = \begin{cases} x, & \text{if } x \in \mathbb{Q}, \\ -x, & \text{if } x \notin \mathbb{Q}. \end{cases}$$

Excercise 8.5 Let $f, g : [a, b] \to \mathbb{R}$ be Riemann integrable. If $g \neq 0$ and $\frac{1}{g}$ is bounded, then prove that $\frac{f}{g} : [a, b] \to \mathbb{R}$ is Riemann integrable.

Excercise 8.6 If f is Riemann integrable on $[a, b]$, then prove that f^2 is also Riemann integrable on $[a, b]$.

Excercise 8.7 Construct an example of a Riemann integrable function which is not monotone.

Excercise 8.8 Examine the Riemann integrability of the following function:

$$f(x) = \begin{cases} \sin \frac{1}{x}, & \text{if } 0 < x \leq 1, \\ 0, & \text{if } x = 0, \end{cases}$$

which is Riemann integrable on $[0, 1]$.

Excercise 8.9 If $f : [a, b] \to \mathbb{R}$ is a bounded function with finitely many discontinuities, then prove that f is Riemann integrable.

Exercise 8.10 Construct an example of a Riemann integrable function having infinite number of points of discontinuity.

Exercise 8.11 Prove that a bounded function f is Riemann integrable on $[a, b]$ if the set of its points of discontinuity has only a finite number of limit points.

Exercise 8.12 Define a function $f : [-1, 1] \to \mathbb{R}$ by

$$f(x) = \begin{cases} c, & \text{a positive constant if } x \neq 0, \\ 0, & \text{if } x = 0. \end{cases}$$

Show that f is Riemann integrable on $[-1, 1]$, and, also, find

$$\int_{-1}^{1} f(x)dx.$$

Hint. The function f has only one point discontinuity.

Exercise 8.13 Give an example to show that continuity of f is needed in the Fundamental Theorem of Calculus I to ensure that F is differentiable.

Hint. Let

$$f(x) = \begin{cases} 1, & \text{if } x \geq 0, \\ 0, & \text{if } x < 0. \end{cases}$$

Then, the function F is not differentiable at $x = 0$ (the point of discontinuity of f), where

$$F(x) = \int_{0}^{x} f(t)dt = \begin{cases} x, & \text{if } x \geq 0, \\ 0, & \text{if } x < 0. \end{cases}$$

Exercise 8.14 Using the Fundamental Theorem of Calculus II, find the following:

$$\int_{1}^{2} x^5 dx.$$

Hint. Consider $f(x) = x^5$ and $F(x) = \frac{x^6}{6}$.

Exercise 8.15 Find the values of c that satisfies the Mean Value Theorem for integral of $x - x^2$ on $[0, 1]$.

Exercise 8.16 Give an example to show that the condition of continuity is necessary in the Mean Value Theorem for the function to assume its mean value in the given interval.

Hint. Consider the following function:

$$f(x) = \begin{cases} 2, & \text{if } 1 \leq x < 2, \\ 4, & \text{if } 2 \leq x \leq 4, \end{cases}$$

on $[1, 4]$. Then, show that

$$\frac{1}{4-1} \int_1^4 f(x) dx = \frac{10}{3},$$

which f fails to assume on $[1, 4]$.

Excercise 8.17 Use the Substitution Theorem to evaluate the following:

(1) $\int_1^3 x \cos(x^2 + 1) dx$;

(2) $\int_0^1 \sin(\pi^2 x) dx$;

(3) $\int_2^5 \frac{5x}{(x^2+1)^3} dx$.

Hint. (1) Take $g(x) = x^2 + 1$, (2) Take $g(x) = \pi^2 x$, (3) Take $g(x) = x^2 + 1$.

Excercise 8.18 Use the Integration by Parts Theorem to evaluate the following:

(1) $\int_0^{\frac{\pi}{4}} x^2 \sin 2x dx$;

(2) $\int_0^1 x^2 e^{3x} dx$;

(3) $\int_1^3 x^2 \ln |x| dx$.

Excercise 8.19 Compute the following improper integral:

$$\int_{-\infty}^{\infty} e^{-|x|} dx.$$

Ans. 2.

Excercise 8.20 Suppose $p > 0$. Show that the following improper integral:

$$\int_1^{\infty} \frac{1}{x^p} dx$$

converges to $\frac{1}{p-1}$ if $p > 1$ and diverges to ∞ if $0 < p \leq 1$.

Hint. We have

$$\int_1^{\infty} \frac{1}{x^p} dx = \lim_{b \to \infty} \int_1^b \frac{1}{x^p} dx = \lim_{b \to \infty} \left(\frac{b^{1-p} - 1}{1 - p} \right).$$

If $p = 1$, then we have

$$\int_1^{\infty} \frac{1}{x} dx = \lim_{b \to \infty} \int_1^b \frac{1}{x} dx = \lim_{b \to \infty} \ln b = \infty.$$

Excercise 8.21 Show that the following improper integral:

$$\int_0^\infty \frac{\sin x}{1+x^2}\,dx$$

converges absolutely.

Hint. Note that

$$\int_0^\infty \frac{\sin x}{1+x^2}\,dx = \int_0^1 \frac{\sin x}{1+x^2}\,dx + \int_1^\infty \frac{\sin x}{1+x^2}\,dx$$

and

$$\left|\frac{\sin x}{1+x^2}\right| \le \frac{1}{x^2} \quad \text{for every } x \ge 1.$$

Excercise 8.22 Show that the following improper integral:

$$\int_0^\infty \frac{\sin x}{x}\,dx$$

converges conditionally.

Excercise 8.23 Let $f(x) = c$ for every $x \in [a,b]$, where c is a constant, and α be a monotonically non-decreasing function on $[a,b]$. Prove that f is Riemann-Stieltjes integrable on $[a,b]$ and

$$\int_a^b c\,d\alpha = c\{\alpha(b) - \alpha(a)\}.$$

Excercise 8.24 Evaluate

$$\int_0^5 x\,d[x],$$

where $[x]$ is the greatest integer not exceeding x.

Excercise 8.25 Evaluate

$$\int_0^1 x\,dx^2,$$

by using the definition of the Riemann-Stieltjes integral.

Ans. $\frac{2}{3}$.

Excercise 8.26 If $f \in \Re_\alpha[a,b]$, then prove that

$$m\{\alpha(b) - \alpha(a)\} \le \int_a^b f\,d\alpha \le M\{\alpha(b) - \alpha(a)\},$$

where m and M are the bounds of f on $[a,b]$.

Excercise 8.27 Prove the following:

(1) If $f \in \Re_\alpha[a, b]$, then there exists a number k lying between the bounds of f such that

$$\int_a^b f d\alpha = k\{\alpha(b) - \alpha(a)\};$$

(2) If f is continuous on $[a, b]$, then there exists a number c lying between a and b such that

$$\int_a^b f d\alpha = f(c)\{\alpha(b) - \alpha(a)\};$$

(3) If $f \in \Re_\alpha[a, b]$ and $f(x) \geq 0$ for every $x \in [a, b]$, then

$$\int_a^b f d\alpha \geq 0;$$

(4) If $f, g \in \Re_\alpha[a, b]$ and $f(x) \leq g(x)$ on $[a, b]$, then we have

$$\int_a^b f d\alpha \leq \int_a^b g d\alpha.$$

Hint. Use Exercise 8.26.

Excercise 8.28 Let $f, g \in \Re_\alpha[a, b]$ and k be a constant. Prove the following:

(1) $f + g \in \Re_\alpha[a, b]$ and

$$\int_a^b (f + g) d\alpha = \int_a^b f d\alpha + \int_a^b g d\alpha;$$

(2) $kf \in \Re_\alpha[a, b]$ and

$$\int_a^b k f d\alpha = k \int_a^b f d\alpha;$$

(3) If $a < c < b$, then $f \in \Re_\alpha[a, c]$, $f \in \Re_\alpha[c, b]$ and

$$\int_a^c f d\alpha + \int_c^b f d\alpha = \int_a^b f d\alpha;$$

(4) $|f| \in \Re_\alpha[a, b]$ and

$$\left| \int_a^b f d\alpha \right| \leq \int_a^b |f| d\alpha;$$

(5) $f^2 \in \Re_\alpha[a, b]$ and $fg \in \Re_\alpha[a, b]$.

Hint. The proofs can be obtained using similar arguments as applied to the corresponding theorems of Riemann integral.

Excercise 8.29 Prove the following:

(1) If $f \in \Re_{\alpha_1}[a, b]$ and $f \in \Re_{\alpha_2}[a, b]$, then $f \in \Re_{(\alpha_1 + \alpha_2)}[a, b]$ and

$$\int_a^b f d(\alpha_1 + \alpha_2) = \int_a^b f d\alpha_1 + \int_a^b f d\alpha_2;$$

(2) If $f \in \Re_{\alpha}[a, b]$ and k is a constant, then $f \in \Re_{(k\alpha)}[a, b]$ and

$$\int_a^b f d(k\alpha) = k \int_a^b f d\alpha.$$

Excercise 8.30 Evaluate the Riemann-Stieltjes integral

$$\int_0^{\frac{\pi}{4}} (x + 1) d(\cos x + \sin x).$$

Hint. Use Exercise 8.28(1) and Exercise 8.29(1).

Excercise 8.31 If α and β are monotonically increasing on $[a, b]$, then show that $\alpha - \beta$ is of bounded variation on $[a, b]$. Is the converse true?

Hint. See Theorem 8.7.5 for the converse part.

Excercise 8.32 Prove the following:

(1) If $\alpha, \beta \in BV[a, b]$, then $\alpha\beta \in BV[a, b]$;
(2) If $\alpha \in BV[a, b]$ and if $|\alpha(x)| \geq c$ on $[a, b]$ for some $c > 0$, then $\frac{1}{\alpha} \in BV[a, b]$;
(3) If $\alpha, \beta \in BV[a, b]$ and $\frac{1}{\beta}$ is bounded on $[a, b]$, then $\frac{\alpha}{\beta} \in BV[a, b]$.

Excercise 8.33 Show the following:

(1) $\sin x$ is of bounded variation on a finite interval;
(2) Every polynomial function is of bounded variation on a finite interval.

Excercise 8.34 Show that the following function

$$f(x) = \begin{cases} 1, & \text{if } x \in \mathbb{Q}, \\ 0, & \text{if } x \notin \mathbb{Q}, \end{cases}$$

is not of bounded variation on any interval.

Excercise 8.35 Determine whether or not α and β are of bounded variation on $[-1, 1]$, where

$$\alpha(x) = \begin{cases} x^2 \sin\left(\frac{1}{x}\right), & \text{if } x \neq 0, \\ 0, & \text{if } x = 0, \end{cases}$$

and

$$\beta(x) = \begin{cases} x^2 \sin\left(\frac{1}{x^2}\right), & \text{if } x \neq 0, \\ 0, & \text{if } x = 0. \end{cases}$$

9

Sequences and Series of Functions

In this chapter, we introduce the pointwise convergence of sequences of functions, the Cauchy Criterion of the uniform convergence, the uniform convergence of infinite series of functions, the uniform convergence of integrations, the uniform convergence of differentiations, the equicontinuous family of functions, the Arzela-Ascoli theorem, Dirichlet's test for uniform convergence and the Weierstrass theorem, and others.

9.1 The Pointwise Convergence of Sequences of Functions and the Uniform Convergence

In this section, we deal with the sequences $\{f_n\}$, whose terms are real- or complex-valued functions having a common domain in $\mathbb{R}$, the field of real numbers or $\mathbb{C}$, and the field of complex numbers. For every x in the domain, consider the sequence $\{f_n(x)\}$. Let S be the set of all x for which $\{f_n(x)\}$ converges. We define the function f by

$$f(x) = \lim_{n \to \infty} f_n(x) \text{ for every } x \in S.$$

Then, the function f is called the *limit function* of $\{f_n\}$. We say that $\{f_n\}$ converges to f *pointwise* on the set S.

In this section, we are concerned about the following question:

If each term of $\{f_n\}$ has a certain property, such as the continuity, the differentiability, or the integrability, can this properly be transferred to the limit function f?

For instance, if each f_n is continuous at a point c, is the limit function also continuous at c?

We formulate this question in another way: Does

$$\lim_{x \to c} f_n(x) = f_n(c) \text{ for every } n = 1, 2, \ldots$$

imply

$$\lim_{x \to c} f(x) = f(c). \tag{9.1}$$

We note that, if yes, then (9.1) can be written as follows:

$$\lim_{x \to c} \lim_{n \to \infty} f_n(x) = \lim_{n \to \infty} \lim_{x \to c} f_n(x). \tag{9.2}$$

Thus, the question of continuity amounts to the following:

Can we interchange the limits in (9.2)?

One can prove that we cannot do it in general. To do this, we need stronger forms of the convergence that preserve properties such as the continuity. With this motivation, we introduce the concept of the "uniform convergence". To make the study complete, we now give an example of a sequence of continuous functions with a discontinuous limit function as follows:

Example 9.1.1. Let

$$f_n(x) = \frac{x^{2n}}{1 + x^{2n}} \quad \text{for every } x \in \mathbb{R}, \, n = 1, 2, \dots.$$

Note that $\lim_{n \to \infty} f_n(x)$ exists for every $x \in \mathbb{R}$ and the limit function f is given by

$$f(x) = \begin{cases} 0, & \text{if } |x| < 1, \\ \frac{1}{2}, & \text{if } |x| = 1, \\ 1, & \text{if } |x| > 1. \end{cases}$$

We also note that f_n is continuous on $\mathbb{R}$, for every $n = 1, 2, \dots$. However, f is discontinuous at $x = \pm 1$.

Let $\{f_n\}$ be a sequence of functions converging to a limit function pointwise on a set S. Then, for every $\epsilon > 0$, there exists a positive integer N, depending both on ϵ and x, such that

$$|f_n(x) - f(x)| < \epsilon \quad \text{whenever } n > N.$$

If the same N works equally well for every $x \in S$, then the convergence is said to be *uniform* on S. Thus, we have the following definition:

Definition 9.1.1. A sequence of functions $\{f_n\}$ is said to *converge uniformly* to a function f on a set S if, for every $\epsilon > 0$, there exists a positive integer N, depending only on ϵ such that

$$|f_n(x) - f(x)| < \epsilon \quad \text{for every } n > N, \, x \in S. \tag{9.3}$$

Symbolically, it is written as follows:

$$f_n \to f \quad \text{as } n \to \infty \quad \text{uniformly on } S$$

or

$$\lim_{n \to \infty} f_n = f \quad \text{uniformly on } S.$$

When each term of $\{f_n\}$ is real-valued, we have a useful geometric interpretation of the uniform convergence. In this case, (9.3) means that

$$f(x) - \epsilon < f_n(x) < f(x) + \epsilon \text{ for every } n > N, x \in S.$$

Consequently, the entire graph of f_n, i.e., $\{(x, y) : y = f_n(x), x \in S\}$ lies within a band of height 2ϵ placed symmetrically about the graph of f.

Definition 9.1.2. A sequence $\{f_n\}$ is said to be *uniformly bounded* if there exists $M > 0$ such that

$$|f_n(x)| \leq M \text{ for every } n = 1, 2, \ldots, x \in S. \tag{9.4}$$

The member M is called a *uniform bound* of $\{f_n\}$.

Theorem 9.1.1. *If $f_n \to f$ as $n \to \infty$ uniformly on S and each f_n is bounded, then $\{f_n\}$ is uniformly bounded on S.*

Proof. Since $f_n \to f$ as $n \to \infty$ uniformly on S, (9.3) holds, i.e.,

$$|f_n(x) - f(x)| < \epsilon \text{ for every } n > N, x \in S.$$

Now, for every $n > N$ and $x \in S$, we have

$$\begin{aligned} |f_n(x)| &= |\{f_n(x) - f(x)\} + f(x)| \\ &\leq |f_n(x) - f(x)| + |f(x)| \\ &< |f(x)| + \epsilon. \end{aligned}$$

For every $n = 1, 2, \ldots, N$, f_n is bounded on S so that there exists $M(n) > 0$ such that

$$|f_n(x)| \leq M(n) \text{ for every } x \in S.$$

Let $M = \max\limits_{1 \leq n \leq N} M(n)$. Then, for every $n = 1, 2, \ldots$ and $x \in S$, we have

$$|f_n(x)| \leq \max\{M, |f(x)| + \epsilon\}.$$

In other words, $\{f_n\}$ is uniformly bounded on S. This completes the proof.

9.2 The Uniform Convergence and the Continuity, the Cauchy Criterion for the Uniform Convergence

Now, we prove the following important result:

Theorem 9.2.1. *Let $f_n \to f$ as $n \to \infty$ uniformly on S. If f_n is continuous at a point $c \in S$ for every $n = 1, 2, \ldots$, then the limit function f is also continuous at the point c.*

Proof. If c is an isolated point of S, the conclusion is clear.

Let us now suppose that c is an accumulation point of S. By the hypothesis, for every $\epsilon > 0$, there exists a positive integer N such that

$$|f_n(x) - f(x)| < \frac{\epsilon}{3} \text{ for every } n \geq N, \, x \in S. \tag{9.5}$$

Since f_N is continuous at the point c, there exists a neighborhood $B(c)$ of c such that

$$|f_N(x) - f_N(c)| < \frac{\epsilon}{3} \text{ whenever } x \in B(c) \bigcap S. \tag{9.6}$$

Now, note that, in view of (9.5) and (9.6),

$$\begin{aligned} |f(x) - f(c)| &= |\{f(x) - f_N(x)\} + \{f_N(x) - f_N(c)\} + \{f_N(c) - f(c)\}| \\ &\leq |f(x) - f_N(x)| + |f_N(x) - f_N(c)| + |f_N(c) - f(c)| \\ &< \frac{\epsilon}{3} + \frac{\epsilon}{3} + \frac{\epsilon}{3} \\ &= \epsilon, \end{aligned}$$

whenever $x \in B(c) \bigcap S$, i.e., $|f(x) - f(c)| < \epsilon$ whenever $x \in B(c) \cap S$, i.e., f is continuous at c. This completes the proof.

Definition 9.2.1. (The Cauchy Criterion) Let $\{f_n\}$ be a sequence of functions defined on a set S. $\{f_n\}$ is said to satisfy the *Cauchy Criterion* if, for every $\epsilon > 0$, there exists a positive integer N such that

$$|f_m(x) - f_n(x)| < \epsilon \text{ for every } m, n > N, \, x \in S. \tag{9.7}$$

Theorem 9.2.2. *Let $\{f_n\}$ be a sequence of functions defined on S. Then, the following are equivalent:*

(1) *There exists a function f such that $f_n \to f$ as $n \to \infty$ uniformly on S;*

(2) *The Cauchy Criterion is satisfied, i.e., (9.7) is fulfilled.*

Proof. (1) $\implies$ (2) Let $f_n \to f$ as $n \to \infty$ uniformly on S. Then, for every $\epsilon > 0$, there exists a positive integer N such that

$$|f_n(x) - f(x)| < \frac{\epsilon}{2} \text{ for every } n > N, \, x \in S.$$

We also have

$$|f_m(x) - f(x)| < \frac{\epsilon}{2} \text{ for every } m > N, \, x \in S.$$

Consequently, we have

$$\begin{aligned} |f_m(x) - f_n(x)| &= |\{f_m(x) - f(x)\} - \{f_n(x) - f(x)\}| \\ &\leq |f_m(x) - f(x)| + |f_n(x) - f(x)| \\ &< \frac{\epsilon}{2} + \frac{\epsilon}{2} \\ &= \epsilon \end{aligned}$$

for every $m, n > N$ and $x \in S$.

(2) $\implies$ (1) Let the Cauchy Criterion be fulfilled. Note that the sequence $\{f_n(x)\}$ converges for every $x \in S$. Let

$$\lim_{n \to \infty} f_n(z) = f(x) \quad \text{for every } x \in S.$$

Now, we claim that

$$f_n \to f \quad \text{as} \quad n \to \infty \quad \text{uniformly on } S.$$

By the hypothesis, for every $\epsilon > 0$, there exists a positive integer N such that

$$|f_n(x) - f_m(x)| < \frac{\epsilon}{2} \quad \text{for every } m, n > N, \, x \in S. \tag{9.8}$$

Taking the limit as $m \to \infty$ in (9.8), we have

$$|f_n(x) - f_m(x)| \leq \frac{\epsilon}{2} < \epsilon \quad \text{for every } n > N, \, x \in S,$$

i.e., $f_n \to f$ as $n \to \infty$ uniformly on S. This is our claim. This completes the proof.

9.3 The Uniform Convergence of Infinite Series of Functions

Definition 9.3.1. Let $\{f_n\}$ be a sequence of functions defined on a set S. For every $x \in S$, let

$$s_n(x) = \sum_{k=1}^{n} f_k(x) \quad \text{for every } n = 1, 2, \ldots. \tag{9.9}$$

If there exists a function f such that $s_n \to f$ as $n \to \infty$ uniformly on S, we say that the series $\sum_{n=1}^{\infty} f_n(x)$ *converges uniformly* on S. We write

$$\sum_{n=1}^{\infty} f_n(x) = f(x) \quad \text{uniformly on } S.$$

Defining a sequence $\{s_n\}$, using (9.9) and applying Theorem 9.2.1, we have the following:

Theorem 9.3.1. (The Cauchy Criterion for the Uniform Convergence of Series) *The following are equivalent:*

(1) *The infinite series $\sum_{n=1}^{\infty} f_n(x)$ converges uniformly on a set S;*

(2) *For every $\epsilon > 0$, there exists a positive integer N such that*

$$\left|\sum_{k=n+1}^{n+p} f_k(x)\right| < \epsilon \ \text{ for every } \ n > N, \ x \in S, \ p = 1, 2, \ldots. \tag{9.10}$$

Theorem 9.3.2. (The Weierstrass M-Test) *Let* $\{M_n\}$ *be a sequence of non-negative numbers such that*

$$0 \le |f_n(x)| \le M_n \ \text{ for every } x \in S, \ n = 1, 2, \ldots.$$

If $\sum_{n=1}^{\infty} M_n$ *converges, then* $\sum_{n=1}^{\infty} f_n(x)$ *converges uniformly on* S.

Proof. First, we note that

$$\left|\sum_{k=n+1}^{n+p} f_k(x)\right| \le \sum_{k=n+1}^{n+p} |f_k(x)| \le \sum_{k=n+1}^{n+p} M_k.$$

Since $\sum_{n=1}^{\infty} M_n$ converges, for every $\epsilon > 0$, there exists a positive integer N such that

$$\sum_{k=n+1}^{n+p} M_k < \epsilon \ \text{ for every } n > N, \ p = 1, 2, \ldots.$$

Thus, we have

$$\left|\sum_{k=n+1}^{n+p} f_k(x)\right| < \epsilon \ \text{ for every } \ n > N, \ x \in S, \ p = 1, 2, \ldots,$$

i.e., $\sum_{n=1}^{\infty} f_n(x)$ converges uniformly on S in view of Theorem 9.3.1. This completes the proof.

Theorem 9.3.3. *Let*

$$\sum_{n=1}^{\infty} f_n(x) = f(x) \ \text{ uniformly on } \ S.$$

If each f_n *is continuous at a point* $c \in S$, *then* f *is also continuous at the point* c.

Proof. Define a sequence $\{s_n\}$ using (9.9). The continuity of each f_n at the point c implies the continuity of each $s_n = f_1 + f_2 + \cdots + f_n$ at c. The conclusion immediately follows using Theorem 9.2.1.

9.4 The Uniform Convergence of Integrations and Differentiations

First, we prove an important result concerning the uniform convergence of integrations.

Theorem 9.4.1. *Let α be monotonically increasing on $[a, b]$. Let $f_n \in \Re(\alpha)$ on $[a, b]$ for every $n = 1, 2, \ldots$. If $f_n \to f$ as $n \to \infty$ uniformly on $[a, b]$, then $f \in \Re(\alpha)$ on $[a, b]$. Further, $\lim_{n \to \infty} \int_a^b f_n d\alpha$ exists and*

$$\int_a^b f d\alpha = \lim_{n \to \infty} \int_a^b f_n d\alpha.$$

Proof. It suffices to prove the theorem when f_n is real valued for every $n = 1, 2, \ldots$. Let

$$\epsilon_n = \sup_{a \leq x \leq b} |f_n(x) - f(x)| \quad \text{for every } n = 1, 2, \ldots.$$

Then, we have

$$f_n - \epsilon_n \leq f \leq f_n + \epsilon_n,$$

so that

$$\int_a^b (f_n - \epsilon_n) d\alpha \leq \underline{\int_a^b} f d\alpha \leq \overline{\int_a^b} f d\alpha \leq \int_a^b (f_n + \epsilon_n) d\alpha. \tag{9.11}$$

Now, using (9.11), we have

$$0 \leq \overline{\int_a^b} f d\alpha - \underline{\int_a^b} f d\alpha \tag{9.12}$$

$$\leq \int_a^b (f_n + \epsilon_n) d\alpha - \int_a^b (f_n - \epsilon_n) d\alpha$$

$$= 2\epsilon_n [\alpha(b) - \alpha(a)]. \tag{9.13}$$

Since $f_n \to f$ as $n \to \infty$ uniformly on $[a, b]$, for every $\epsilon > 0$, there exists a positive integer N such that

$$|f_n(x) - f(x)| < \epsilon \quad \text{for every } n > N, \, x \in [a, b].$$

So, we have

$$\epsilon_n = \sup_{a \leq x \leq b} |f_n(x) - f(x)| \leq \epsilon \quad \text{for every } n > N,$$

which implies that

$$\epsilon_n \to 0 \text{ as } n \to \infty.$$

It now follows from (9.12) that

$$\underline{\int_a^b} f\,d\alpha = \overline{\int_a^b} f\,d\alpha.$$

Thus, $f \in \Re(\alpha)$. Taking limit as $n \to \infty$ in (9.11), we see that $\lim_{n\to\infty} \int_a^b f_n\,d\alpha$ exists and

$$\lim_{n\to\infty} \int_a^b f_n\,d\alpha = \int_a^b f\,d\alpha.$$

This completes the proof.

As a consequence of Theorem 9.4, we have the following:

Corollary 9.4.2. *If $f_n \in \Re(\alpha)$ on $[a, b]$, where α is monotonically increasing on $[a, b]$, and*

$$\sum_{n=1}^{\infty} f_n(x) = f(x) \text{ for every } x \in [a, b],$$

that is, the infinite series converging uniformly on $[a, b]$, then we have

$$\int_a^b f\,d\alpha = \sum_{n=1}^{\infty} \int_a^b f_n\,d\alpha.$$

In other words, the series $\sum_{n=1}^{\infty} f_n(x)$ can be integrated term by term.

Next, we have the following result concerning the uniform convergence of differentiations:

Theorem 9.4.3. *Let $\{f_n\}$ be a sequence of functions each of which is differentiable on $[a, b]$. Suppose that $\{f_n(x_0)\}$ converges for some point $x_0 \in [a, b]$. If $\{f_n'\}$ converges uniformly on $[a, b]$, then $\{f_n\}$ converges uniformly on $[a, b]$ to a function f and*

$$f'(x) = \lim_{n\to\infty} f_n'(x) \text{ for every } x \in [a, b].$$

Proof. Since $\{f_n(x_0)\}$ converges and $\{f_n'\}$ converges uniformly on $[a, b]$, for every $\epsilon > 0$, we can choose a positive integer N such that

$$|f_n(x_0) - f_m(x_0)| < \frac{\epsilon}{2} \text{ for every } n, m > N \tag{9.14}$$

and

$$|f_n'(t) - f_m'(t)| < \frac{\epsilon}{2(b-a)} \text{ for every } t \in [a, b]. \tag{9.15}$$

Applying the Mean Value Theorem to the function $f_n - f_m$ and using (9.15), we have

$$|\{f_n(x) - f_m(x)\} - \{f_n(t) - f_m(t)\}| \leq \frac{|x - t|\epsilon}{2(b - a)} < \frac{(b - a)\epsilon}{2(b - a)} = \frac{\epsilon}{2} \qquad (9.16)$$

for every $x, t \in [a, b]$ and $n, m > N$. Now, in view of (9.14) and (9.16), we have

$$
\begin{aligned}
&|f_n(x) - f_m(x)| \\
&= |\{f_n(x) - f_m(x)\} - \{f_n(x_0) - f_m(x_0)\} + \{f_n(x_0) - f_m(x_0)\}| \\
&\leq |\{f_n(x) - f_m(x)\} - \{f_n(x_0) - f_m(x_0)\}| + |\{f_n(x_0) - f_m(x_0)\}| \\
&< \frac{\epsilon}{2} + \frac{\epsilon}{2} \\
&= \epsilon
\end{aligned}
$$

for every $x \in [a, b]$ and $n, m > N$, i.e., $\{f_n\}$ converges uniformly on $[a, b]$. Let

$$\lim_{n \to \infty} f_n(x) = f(x) \text{ for every } x \in [a, b].$$

Fix a point $x \in [a, b]$ and define

$$\phi_n(t) = \frac{f_n(t) - f_n(x)}{t - x}, \quad \phi(t) = \frac{f(t) - f(x)}{t - x} \text{ for every } t \in [a, b], \, t \neq x. \qquad (9.17)$$

Then, we have

$$\lim_{t \to x} \phi_n(t) = f'_n(x) \text{ for every } n = 1, 2, \ldots. \qquad (9.18)$$

Using (9.16), we have $|\phi_n(t) - \phi_m(t)| < \epsilon$ for every $n, m > N$, $t \in [a, b]$ and $t \neq x$, so that $\{\phi_n\}$ converges uniformly for every $t \neq x$. Since $\{f_n\}$ converges to f, using (9.17), we have

$$\lim_{n \to \infty} \phi_n(t) = \phi(t) \text{ uniformly for every } t \in [a, b], \, t \neq x. \qquad (9.19)$$

Applying Theorem 9.2 to the sequence $\{\phi_n\}$, we have

$$\lim_{t \to x} \lim_{n \to \infty} \phi(t) = \lim_{n \to \infty} \lim_{t \to x} \phi_n(t),$$

i.e., using (9.18) and (9.19), we have

$$\lim_{t \to x} \phi(t) = \lim_{n \to \infty} f'_n(x),$$

i.e., using the definition of $\phi(t)$, we have

$$f'(x) = \lim_{n \to \infty} f'_n(x) \text{ for every } x \in [a, b].$$

This completes the proof.

9.5   The Equicontinuous Family of Functions and the Arzela-Ascoli Theorem

Let X be a compact metric space. The set of all continuous real-valued functions defined on X with the metric

$$\rho(f, g) = \sup_{x \in X} |f(x) - g(x)|$$

is a complete metric space denoted by $C(X)$.

Definition 9.5.1. A collection $\mathscr{F}$ of real-valued functions defined on a set X is said to be *uniformly bounded* if there exists $M > 0$ such that

$$|f(x)| \leq M \quad \text{for every } x \in X, \, f \in \mathscr{F}.$$

Definition 9.5.2. A collection $\mathscr{F}$ of real-valued functions defined on a metric space X is said to be *equicontinuous* if, for every $\epsilon > 0$, there exists $\delta > 0$ such that $\rho(x, x') < \delta$ implies

$$|f(x) - f(x')| < \epsilon \quad \text{for every } x, x' \in X, \, f \in \mathscr{F}.$$

In the context of Definition 9.5, we note that functions belonging to the equicontinuous family are uniformly continuous.

Definition 9.5.3. A metric space X is said to be *totally bounded* if, for every $\epsilon > 0$, X contains a finite set, called an ϵ-*net*, such that the finite set of open spheres with centers in the ϵ-net and radius ϵ covers X.

With this background, we now prove a very useful result.

Theorem 9.5.1. (The Arzela-Ascoli Theorem) *Let X be a compact metric space. Then, the following are equivalent:*

(1) *A subset K of $C(X)$ is relatively compact;*

(2) *It is uniformly bounded and equicontinuous.*

Proof. (1) $\Longrightarrow$ (2) Suppose that K is relatively compact. Then, K is totally bounded. Let $\epsilon > 0$. Let $\{f_1, f_2, \ldots, f_n\}$ be an $\frac{\epsilon}{3}$-net in K. Now, let $f \in K$ and $x, x' \in X$. For every $i = 1, 2, \ldots, n$, we have

$$|f(x) - f(x')| \leq |f(x) - f_i(x)| + |f_i(x) - f_i(x')| + |f_i(x') - f(x')|.$$

Choose j, $1 \leq j \leq n$ such that

$$\sup_{x \in X} |f(x) - f_j(x)| < \frac{\epsilon}{3},$$

so that

$$|f(x) - f(x')| < |f_j(x) - f_j(x')| + \frac{2\epsilon}{3}. \tag{9.20}$$

Since X is compact, the functions f_i, $i = 1, 2, \ldots, n$ are uniformly continuous. So, there exists $\delta > 0$ such that $\rho(x, x') < \delta$ implies

$$|f_i(x) - f_i(x')| < \frac{\epsilon}{3} \quad \text{for every} \ \ i = 1, 2, \ldots, n. \tag{9.21}$$

In view of (9.20) and (9.21), we have

$$|f(x) - f(x')| < \frac{\epsilon}{3} + \frac{2\epsilon}{3} = \epsilon$$

whenever $\rho(x, x') < \delta$. In other words, K is equicontinuous.

$(2) \implies (1)$ Let K be uniformly bounded and equicontinuous. Let M be an integer such that

$$|f(x)| \le M \quad \text{for every} \ x \in X, \ f \in K.$$

Let $\epsilon > 0$. Choose $\delta > 0$ such that $\rho(x, x') < \delta$ implies

$$|f(x) - f(x')| < \frac{\epsilon}{4} \quad \text{for every} \ f \in K.$$

Since X is compact, it has a δ-net (say) $\{x_1, x_2, \ldots, x_n\}$. Let m be a positive integer such that

$$\frac{1}{m} < \frac{\epsilon}{4}.$$

Now, divide $[-M, M]$ into $2Mm$ equal parts by the points

$$y_0 = -M < y_1 < y_2 < \cdots < y_k = M,$$

where $k = 2Mm$. We shall consider those n-tuples $(y_{i_1}, y_{i_2}, \ldots y_{i_n})$ of the numbers y_i, $i = 0, 1, \ldots, k$, such that some $f \in K$ has the property

$$|f(x_j) - y_{i_j}| < \frac{\epsilon}{4} \quad \text{for every} \ \ j = 1, 2, \ldots, n,$$

and let us choose one such $f \in K$ for each such n-tuple.

Now, we claim that the resulting finite subset E of K is an ϵ-net for K. If $f \in K$, we choose $(y_{i_1}, y_{i_2}, \ldots y_{i_n})$ such that

$$|f(x_j) - y_{i_j}| < \frac{\epsilon}{4} \quad \text{for every} \ \ j = 1, 2, \ldots, n$$

and so there is a corresponding $e = (y_{i_1}, y_{i_2}, \ldots y_{i_n}) \in E$. Let $x \in X$, and choose j such that $\rho(x, x_j) < \delta$. Then, we have

$$|f(x) - e(x)| \le |f(x) - f(x_j)| + |f(x_j) - y_{i_j}| + |y_{i_j} - e(x_j)| + |e(x_j) - e(x)|$$
$$< 4\frac{\epsilon}{4} = \epsilon.$$

Thus, we have

$$\sup_{x \in X} |f(x) - e(x)| \le \epsilon.$$

This completes the proof.

9.6 Dirichlet's Test for the Uniform Convergence

In an attempt to seek some simple ways of testing a series for uniform convergence without resorting to the definition in each case, we had the Weierstrass M-test introduced already. There are other tests that may be useful when the M-test is not applicable. One of these tests is as follows:

Theorem 9.6.1. (Dirichlet's Test for the Uniform Convergence) *Let*

$$F_n(x) = \sum_{k=1}^{n} f_k(x) \ \text{ for every } \ n = 1, 2, \ldots,$$

where each f_n is a complex-valued function defined on a S. Let $\{F_n\}$ be uniformly bounded on S. Let $\{g_n\}$ be a sequence of real-valued functions such that $g_{n+1}(x) \le g_n(x)$ for every $x \in S$ and $n = 1, 2, \ldots$. Let $g_n \to 0$ as $n \to \infty$ uniformly on S. Then, the series $\sum_{n=1}^{\infty} f_n(x)g_n(x)$ converges uniformly on S.

Proof. Let

$$s_n(x) = \sum_{k=1}^{n} f_k(x)g_k(x) \ \text{ for every } \ n = 1, 2, \ldots, \, x \in S.$$

By the partial summation, we have

$$s_n(x) = \sum_{k=1}^{n} F_k(x)\{g_k(x) - g_{k+1}(x)\} + g_{n+1}(x)F_n(x).$$

Hence, for every $n > m$, we can write as follows:

$$s_n(x) - s_m(x)$$

$$= \sum_{k=m+1}^{n} F_k(x)\{g_k(x) - g_{k+1}(x)\} + g_{n+1}(x)F_n(x) - g_{m+1}(x)F_m(x).$$

Since $\{F_n\}$ is uniformly bounded on S, there exists $M > 0$ such that

$$|F_n(x)| \le M \ \text{ for all } \ x \in S, \, n = 1, 2, \ldots.$$

Thus, we have

$$|s_n(x) - s_m(x)| \le M \sum_{k=m+1}^{n} \{g_k(x) - g_{k+1}(x)\} + Mg_{n+1}(x) + Mg_{m+1}(x)$$

$$= M\{g_{m+1}(x) - g_{n+1}(x)\} + Mg_{n+1}(x) + Mg_{m+1}(x)$$

$$= 2Mg_{m+1}(x).$$

Since $g_n \to 0$ as $n \to \infty$ uniformly on S, for every $\epsilon > 0$, we can find a positive integer N such that

$$|g_n(x)| < \frac{\epsilon}{2M} \ \text{ for every } \ n > N, \, x \in S.$$

Thus, for every $n > m$, we have

$$|s_n(x) - s_m(x)| < 2M \left[\frac{\epsilon}{2M} \right] = \epsilon \text{ for every } n, m > N, \ x \in S.$$

In other words, using the Cauchy Criterion for the uniform convergence, $\sum_{n=1}^{\infty} f_n(x) g_n(x)$ converges uniformly on S. This completes the proof.

Remark 9.6.1. In the context of Theorem 9.6.1, we note that there is another test to prove the uniform convergence of a series, due to Abel, which is given as Exercise 9.14 toward the end of the chapter.

9.7 The Weierstrass Theorem

In this section, we prove an important result due to Weierstrass on the uniform approximation of a continuous function by a sequence of polynomials.

Theorem 9.7.1. *Let f be a continuous complex-valued function on $[a, b]$. Then, there exists a sequence $\{P_n\}$ of polynomials such that*

$$\lim_{n \to \infty} P_n(x) = f(x) \quad \text{uniformly on} \ \ [a, b].$$

If f is a real-valued function on $[a, b]$, than P_n may be taken as a real-valued function on $[a, b]$.

Proof. Without loss of generality, we can suppose that $[a, b] = [0, 1]$. We may also assume that $f(0) = f(1) = 0$. In fact, if the theorem is proved in the above case, let

$$g(x) = f(x) - f(0) - x[f(1) - f(0)] \quad \text{for every} \ \ x \in [0, 1].$$

Then, we have $g(0) = g(1) = 0$ and, if g can be obtained as the limit of a uniformly convergent sequence of polynomials, then it clear that the same is valid for f too since $f - g$ is a polynomial. Further, we define

$$f(x) = 0 \quad \text{outside} \ [0, 1].$$

Then, f is uniformly continuous on the whole line. Let

$$Q_n(x) = c_n(1 - x^2)^n \quad \text{for every} \ \ n = 1, 2, \ldots, \tag{9.22}$$

where we choose c_n such that

$$\int_{-1}^{1} Q_n(x) dx = 1 \quad \text{for every} \ \ n = 1, 2, \ldots. \tag{9.23}$$

Now, we have

$$\int_{-1}^{1}(1-x^2)^n dx = 2\int_{0}^{1}(1-x^2)^n dx$$

$$\geq 2\int_{0}^{\frac{1}{\sqrt{n}}}(1-x^2)^n dx$$

$$\geq 2\int_{0}^{\frac{1}{\sqrt{n}}}(1-nx^2)dx$$

$$= \frac{4}{3\sqrt{n}} > \frac{1}{\sqrt{n}},$$

from which it follows that

$$c_n < \sqrt{n} \tag{9.24}$$

using (9.22) and (9.23), where note that it is a simple exercise for the reader to prove that

$$\int_{0}^{\frac{1}{\sqrt{n}}}(1-nx^2)dx = \frac{2}{3\sqrt{n}}.$$

In the above working, we also used the fact that

$$(1-x^2)^n \geq 1-nx^2,$$

which is again left as an exercise to the reader (consider the function $(1-x^2)^n - 1 + nx^2$, which is zero at $x = 0$ and whose derivative is positive in $(0,1)$).

For every $\delta > 0$, (9.24) implies that

$$Q_n(x) \leq \sqrt{n}(1-\delta^2)^n \quad \text{for every} \quad \delta \leq |x| \leq 1, \tag{9.25}$$

so that $Q_n \to 0$ as $n \to \infty$ uniformly on $\delta \leq |x| \leq 1$. For every $x \in [0,1]$ and $n = 1, 2, \ldots$, let $x + t = t'$ and then

$$P_n(x) = \int_{-1}^{1} f(x+t)Q_n(t)dt \tag{9.26}$$

$$= \int_{x-1}^{x+1} f(t')Q_n(t'-x)dt'$$

$$= \int_{x-1}^{x+1} f(t)Q_n(t-x)dt$$

$$= \int_{0}^{1} f(t)Q_n(t-x)dt$$

since $f(x) = 0$ outside $[0,1]$. It now follows that P_n is a polynomial in x. Consequently, $\{P_n\}$ is a sequence of polynomials in x, which are real-valued if f is real valued.

Now, for every $\epsilon > 0$, we can choose $\delta > 0$ such that

$$|f(y) - f(x)| < \frac{\epsilon}{2}$$

whenever $|y - x| < \delta$. Let $M = \sup \ \{|f(x)| : x \in [0,1]\}$. Using (9.23), (9.25), and the fact that $Q_n(x) \geq 0$ for every $x \in [0,1]$, it follows that

$$|P_n(x) - f(x)| = \left| \int_{-1}^{1} f(x+t)Q_n(t)dt - \int_{-1}^{1} f(x)Q_n(t)dt \right|$$

$$= \left| \int_{-1}^{1} [f(x+t) - f(x)]Q_n(t)dt \right|$$

$$\leq \int_{-1}^{1} |f(x+t) - f(x)|Q_n(t)dt$$

$$= \int_{-1}^{-\delta} |f(x+t) - f(x)|Q_n(t)dt$$

$$+ \int_{-\delta}^{\delta} |f(x+t) - f(x)|Q_n(t)dt$$

$$+ \int_{\delta}^{1} |f(x+t) - f(x)|Q_n(t)dt$$

$$\leq 2M \int_{-1}^{-\delta} Q_n(t)dt + \frac{\epsilon}{2} \int_{-\delta}^{\delta} Q_n(t)dt + 2M \int_{\delta}^{1} Q_n(t)dt$$

$$\leq 4M\sqrt{n}(1-\delta^2)^n + \frac{\epsilon}{2}$$

$$< \epsilon$$

for sufficiently large n. This completes the proof.

9.8 Some Examples

In this section, we give some examples related to some important theorems in previous sections.

Example 9.8.1. Prove that the operations of the "limit" and the "integration" cannot always be interchanged.

Solution. To prove this statement, we give an example of a sequence of functions $\{f_n\}$ for which

$$\lim_{n \to \infty} \int_0^1 f_n(x)dx \neq \int_0^1 \lim_{n \to \infty} f_n(x)dx.$$

Define

$$f_n(x) = n^2 x(1-x)^n \quad \text{for every } x \in \mathbb{R}, \ n = 1, 2, \ldots.$$

If $x \in [0,1]$, then we have

$$\lim_{n \to \infty} f_n(x) \text{ exists and is equal to } 0,$$

so that

$$\int_0^1 \lim_{n \to \infty} f_n(x) = 0.$$

However, if $n \to \infty$, then we have

$$\int_0^1 f_n(x)dx = n^2 \int_0^1 x(1-x)^n dx$$

$$= n^2 \int_0^1 (1-x)x^n dx$$

$$= n^2 \int_0^1 (x^n - x^{n+1})dx$$

$$= n^2 \left[\frac{x^{n+1}}{n+1} - \frac{x^{n+2}}{n+2} \right]_{x=0}^1$$

$$= n^2 \left[\frac{1}{n+1} - \frac{1}{n+2} \right]$$

$$= \frac{n^2}{(n+1)(n+2)},$$

so that

$$\lim_{n \to \infty} \int_0^1 f_n(x)dx = 1.$$

Consequently, we have

$$\lim_{n \to \infty} \int_0^1 f_n(x)dx \neq \int_0^1 \lim_{n \to \infty} f_n(x)dx.$$

Example 9.8.2. Give an example of a sequence of differentiable functions $\{f_n\}$ with the limit 0 such that $\{f_n'\}$ diverges.

Solution. Define

$$f_n(x) = \frac{\sin nx}{\sqrt{n}} \quad \text{for every} \quad x \in \mathbb{R}, \, n = 1, 2, \ldots.$$

Then, we have

$$|f_n(x)| = \left| \frac{\sin nx}{\sqrt{n}} \right| < \frac{1}{\sqrt{n}} \to 0 \quad \text{as} \quad n \to \infty,$$

so that

$$\lim_{n \to \infty} f_n(x) = 0 \quad \text{for every} \quad x \in \mathbb{R}.$$

However, we have

$$f'_n(x) = \frac{1}{\sqrt{n}} \cos nx \cdot n = \sqrt{n} \cos nx,$$

so that

$$\lim_{n \to \infty} f'_n(x)$$

does not exist for every $x \in \mathbb{R}$.

Example 9.8.3. Prove that the uniform convergence is not a necessary condition for term by term integration.

Solution. Define

$$f_n(x) = x^n \quad \text{for every} \ x \in [0, 1].$$

Note that the limit function f exists and is defined by

$$f(x) = \begin{cases} 0, & \text{if } x \in [0, 1) \\ 1, & \text{if } x = 1, \end{cases}$$

and

$$\int_0^1 f(x)dx = 0.$$

Then, $\{f_n\}$ is a sequence of continuous functions with a discontinuous limit, so that the convergence is not uniform on $[0, 1]$ in view of Theorem 9.2. Now, we have

$$\int_0^1 f_n(x)dx = \int_0^1 x^n dx = \left[\frac{x^{n+1}}{n+1}\right]_{x=0}^1 = \frac{1}{n+1} \to 0 \quad \text{as} \ n \to \infty,$$

so that

$$\lim_{n \to \infty} \int_0^1 f_n(x)dx = 0.$$

Thus, we have

$$\lim_{n \to \infty} \int_0^1 f_n(x)dx = 0 = \int_0^1 f(x)dx = \int_0^1 \lim_{n \to \infty} f_n(x)dx.$$

This proves our claim.

Example 9.8.4. Define

$$f_n(x) = nx(1 - x^2)^n \quad \text{for every} \ x \in [0, 1], \ n = 1, 2, \ldots.$$

Prove that

$$\lim_{n \to \infty} \int_0^1 f_n(x)dx \neq \int_0^1 \lim_{n \to \infty} f_n(x)dx.$$

Solution. For every $0 < x \leq 1$, we have

$$\lim_{n \to \infty} f_n(x) = 0.$$

Since $f_n(0) = 0$, we note that

$$\lim_{n \to \infty} f_n(x) = 0 \quad \text{for every} \quad x \in [0, 1].$$

Now, we have

$$\lim_{n \to \infty} \int_0^1 f_n(x)dx = \lim_{n \to \infty} \int_0^1 nx(1 - x^2)^n dx$$

$$= \lim_{n \to \infty} \frac{n}{2} \int_0^1 (1 - t)^n dt, \ t = x^2$$

$$= \lim_{n \to \infty} \frac{n}{2} \left[\frac{(1 - t)^{n+1}}{-(n+1)} \right]_{t=0}^1$$

$$= \lim_{n \to \infty} -\frac{n}{2(n+1)}[0 - 1]$$

$$= \lim_{n \to \infty} \frac{n}{2(n+1)}$$

$$= \frac{1}{2},$$

while

$$\int_0^1 \lim_{n \to \infty} f_n(x)dx = 0.$$

Thus, we have

$$\lim_{n \to \infty} \int_0^1 f_n(x)dx \neq \int_0^1 \lim_{n \to \infty} f_n(x)dx.$$

Example 9.8.5. Define

$$f_n(x) = \frac{x^2}{x^2 + (1 - nx)^2} \quad \text{for every} \quad x \in [0, 1], \ n = 1, 2, \ldots.$$

Prove that $\{f_n\}$ is not equicontinuous.

Solution. First, note that

$$|f_n(x)| \leq 1 \quad \text{for every} \quad x \in [0, 1], \ n = 1, 2, \ldots,$$

so that $\{f_n\}$ is uniformly bounded on $[0, 1]$. Also, we have

$$\lim_{n \to \infty} f_n(x) = 0 \quad \text{for every} \quad x \in [0, 1].$$

However, we have

$$f_n \left(\frac{1}{n} \right) = 1 \quad \text{for every} \quad n = 1, 2, \ldots$$

and so no any subsequence of $\{f_n\}$ can converge uniformly on $[0, 1]$. It now follows that the sequence $\{f_n\}$ is not equicontinuous.

Example 9.8.6. Define

$$f_n(x) = nxe^{-nx^2} \quad \text{for every } x \in [0,1], \ n = 1, 2, \ldots.$$

Prove that $\{f_n\}$ is not uniformly convergent on $[0,1]$.

Solution. Note that

$$\lim_{n \to \infty} f_n(x) = 0 \quad \text{for every } x \in [0,1].$$

Thus, the sequence $\{f_n\}$ converges pointwise to the function f defined by

$$f_n(x) = 0 \quad \text{for every } x \in [0,1]$$

and so

$$\int_0^1 \lim_{n \to \infty} f_n(x) = 0.$$

However, we have

$$\int_0^1 f_n(x)dx = \int_0^1 nxe^{-nx^2} dx$$

$$= \frac{1}{2} \int_0^n e^{-t}dt, \ t = nx^2$$

$$= \frac{1}{2}[-e^{-t}]_{t=0}^n$$

$$= \frac{1}{2}[1 - e^n]$$

and so

$$\lim_{n \to \infty} \int_0^1 f_n(x)dx = \frac{1}{2}.$$

Thus, we have

$$\lim_{n \to \infty} \int_0^1 f_n(x)dx \neq \int_0^1 \lim_{n \to \infty} f_n(x)dx.$$

Using Theorem 9.4.1, it follows that $\{f_n\}$ is not uniformly convergent on $[0,1]$.

Example 9.8.7. Define

$$f_n(x) = \frac{nx}{1 + n^2 x^2} \quad \text{for every } x \in [0,1], \ n = 1, 2, \ldots.$$

Prove that

$$\lim_{n \to \infty} \int_0^1 f_n(x)dx = \int_0^1 \lim_{n \to \infty} f_n(x)dx.$$

Solution. It is clear that $\{f_n\}$ converges pointwise to 0 on $[0,1]$, so that

$$\int_0^1 \lim_{n \to \infty} f_n(x)dx = 0.$$

Again, if $t = 1 + n^2x^2$, then we have

$$\int_0^1 f_n(x)dx = \int_0^1 \frac{nx}{1+n^2x^2}dx$$

$$= \frac{1}{2n}\int_1^{1+n^2} \frac{dt}{t}$$

$$= \frac{1}{2n}[\log t]_{t=1}^{1+n^2}$$

$$= \frac{1}{2n}[\log(n^2+1) - \log 1]$$

$$= \frac{1}{2n}\log(n^2+1)$$

since $\log 1 = 0$, from which it follows that

$$\lim_{n\to\infty}\int_0^1 f_n(x)dx = 0.$$

Consequently, we have

$$\lim_{n\to\infty}\int_0^1 f_n(x)dx = \int_0^1 \lim_{n\to\infty} f_n(x)dx.$$

9.9 Exercises

Excercise 9.1 Define

$$f_n(x) = \frac{nx}{1+n^2x^2} \quad \text{for every } x \in [0,\ 1],\ n = 1, 2, \dots.$$

(1) Prove that $\{f_n\}$ is not uniformly convergent on $[0, 1]$;

(2) Prove also that, in the case of $\{f_n\}$, the limit of the sequence of derivatives is not equal to the derivative of the limit of the sequence of functions.

Excercise 9.2 Define

$$f_n(x) = nxe^{-nx^2} \quad \text{for every } x \in [0,1],\ n = 1, 2, \dots.$$

Prove directly that $\{f_n\}$ is not uniformly convergent on $[0, 1]$.

Excercise 9.3 Show that the sequence $\{f_n\}$, where

$$f_n(x) = e^{-nx} \quad \text{for every } x \in [0,1],\ n = 1, 2, \dots,$$

is pointwise convergent but not uniformly convergent on $[0, 1]$.

Excercise 9.4 Prove **Dini's Theorem**: If $\{f_n\}$ is a sequence of real-valued continuous functions converging pointwise to a continuous limit function f on a compact set S and if $f_n(x) \geq f_{n+1}(x)$ for every $x \in S$ and $n = 1, 2, \ldots$, then $f_n \to f$ as $n \to \infty$ uniformly on S. (Hint. See [2], p. 248, [6], p. 118.)

Excercise 9.5 Consider $\sum_{n=0}^{\infty} f_n(x)$, where

$$f_n(x) = \frac{x^2}{(1+x^2)^n} \quad \text{for every } x \in \mathbb{R}, n = 0, 1, 2, \ldots.$$

Prove that $\sum_{n=0}^{\infty} f_n(x)$ is a convergent series of continuous functions with a discontinuous sum.

Excercise 9.6 If $\{f_n\}$ and $\{g_n\}$ converge uniformly on a set E, then prove that $\{f_n + g_n\}$ converges uniformly on E. If, in addition, $\{f_n\}$ and $\{g_n\}$ are sequences of bounded functions, then prove that $\{f_n g_n\}$ converges uniformly on E.

Excercise 9.7 Give examples of the sequences $\{f_n\}$ and $\{g_n\}$ which converge uniformly on some set E such that $\{f_n g_n\}$ converges on the set E but does not converge uniformly on E.

Excercise 9.8 Define

$$f_n(x) = \frac{x}{1 + nx^2} \quad \text{for every } x \in \mathbb{R}, n = 1, 2, \ldots.$$

Prove that $\{f_n\}$ converges uniformly to a function f, and prove also that

$$f'(x) = \lim_{n \to \infty} f_n'(x)$$

is valid if $x \neq 0$ and false if $x = 0$.

Excercise 9.9 Let $\{f_n\}$ be a sequence of continuous functions converging uniformly to a function f on a set E. Prove that

$$\lim_{n \to \infty} f_n(x_n) = f(x)$$

for every sequence $\{x_n\}$ of points in E such that $x_n \to x$ as $n \to \infty$, where $x \in E$. Is the converse true? Justify your answer.

Excercise 9.10 Define

$$f_n(x) = \frac{1}{nx + 1} \quad \text{for every } x \in (0, 1), n = 1, 2, \ldots.$$

Prove that $\{f_n\}$ converges pointwise but not uniformly on $(0, 1)$.

Excercise 9.11 Define

$$f_n(x) = \frac{x}{nx+1} \quad \text{for every } x \in (0,1), \ n = 1, 2, \ldots.$$

Prove that $f_n \to 0$ as $n \to \infty$ uniformly on $(0,1)$.

Excercise 9.12 Define

$$f_n(x) = n^\alpha x(1-x^2)^n \quad \text{for every } x \in \mathbb{R}, \ n = 1, 2, \ldots.$$

Prove that $\{f_n\}$ converges pointwise on $[0, 1]$ for every $\alpha \in \mathbb{R}$. Find those α for which the convergence is uniform on $[0, 1]$.

Excercise 9.13 Define

$$f_n(x) = \frac{e^{-n^2 x^2}}{n} \quad \text{for every } x \in \mathbb{R}, \ n = 1, 2, \ldots.$$

Prove that $f_n \to 0$ as $n \to \infty$ uniformly on $\mathbb{R}$ and $f_n' \to 0$ as $n \to \infty$ pointwise on $\mathbb{R}$ but the convergence of $\{f_n'\}$ is not uniform on any interval containing the origin.

Excercise 9.14 (Abel's Test for the Uniform Convergence) Let $\{g_n\}$ be a sequence of real-valued functions such that $g_{n+1}(x) \leq g_n(x)$ for every $x \in E$ and $n = 1, 2, \ldots$. If $\{g_n\}$ is uniformly bounded on E and $\sum_{n=1}^{\infty} f_n(x)$ converges uniformly on E, then prove that $\sum_{n=1}^{\infty} f_n(x)g_n(x)$ also converges uniformly on E.

Bibliography

[1] Q.H. Ansari, Metric *Spaces including Fixed Point Theory and Set-valued Maps*, Alpha Science Internat., Ltd., Oxford, 2010.

[2] T.M. Apostol, *Mathematical Analysis*, Second Edition, Addison-Wesley, Narosa, 1996.

[3] R.G. Bartle and D.R. Sherbert, *Introduction to Real Analysis*, Third Edition, John Wiley & Sons, New York, 2000.

[4] E.T. Copson, *Metric Spaces*, Cambridge University Press, Cambridge, UK, 1968.

[5] L.W. Cohen and G. Ehrlich, *The Structure of the Real Number System*, Van Nostrand, New York, 1963.

[6] V. Ganapathy Iyer, *Mathematical Analysis*, Tata McGraw-Hill Publising Co., New Delhi, 1977.

[7] J.R. Giles, *Introduction to the Analysis of Metric Spaces*, Cambridge University Press, Cambridge, UK, 1987.

[8] C. Goffman and G. Pedrick, *First Course in Functional Analysis*, Prentice, Hall of India Pvt. Ltd., New Delhi, 1974.

[9] R.R. Goldberg, *Methods of Real Analysis*, Second Edition, John Wiley & Sons, Hoboken, NJ, 1976.

[10] P.K. Jain and K. Ahmad, *Metric Spaces*, Narosa Publishing House, Pvt., Ltd., New Delhi, 1993.

[11] W. Rudin, *Principles of Mathematical Analysis*, Third Edition, McGraw-Hill, New York, 1976.

[12] G.F. Simmons, *Introduction to Topology and Modern Analysis*, McGraw-Hill, New York, 1963.

Index

A
Abel's test, 44, 234
Absolute value, real number, 20–21
Algebras, of limits, 63–66
Analytic geometry, 2
Archimedean property, real number system, 16–18
Arzela–Ascoli theorem, 222–223

B
Bolzano–Weierstrass theorem, 51–52, 55, 59, 116
Bounded variation functions, 196–205

C
Cantor's intersection theorem, 52
Cartesian product, definition of, 2–3
Cauchy condition, real numbers, 30–33
Cauchy criterion, 215–217, 225
Cauchy multiplications, 39–41
Cauchy–Schwarz inequality, 22
Cauchy sequences, 25, 54–56
Cauchy's mean value theorem, 137–138, 141, 144
Cesàro's theorem, 43
Chain rule, 128–129
Comparison test
    for innite series, 34
    for integral, 186–187
Complete metric spaces, 54–56
Completeness axiom, real number, 17, 84
Connectedness in metric space

compactness of, 108–113
component of, 107–108
definition of, 99–105
exercises, 120–122
finite intersection property, 114–115
Heine–Borel theorem, 116–120
intermediate value theorem, 105–107
Continuous functions
    definition of, 76
    and discontinuous functions, 75–80
    properties of, 83–85
    theorems on, 80–83
    and uniform continuity, 88–91
Convergence theorems
    Cauchy multiplications of, 39–41
    Cauchy's General Principle of, 43, 73
    convergent and divergent sequences, 25–26
    exercises, 41–44
    infinite series
        convergence tests for, 34–36
        of real numbers, 28–34
    limit superior and limit inferior, 26–28
    rearrangements of, 37–38
    Riemann's theorem, 38–39
Countable sets
    countable collection of, 9
    every subset of, 5–6

Milton Keynes UK
Ingram Content Group UK Ltd.
UKHW040105071024
449327UK00019B/824

9 780367 779283